PROBLEMS IN HEAT AND MASS TRANSFER

22 SEP 1982

D1146905

Problems in Heat and Mass Transfer

J. R. Backhurst, J. H. Harker and J. E. Porter
Department of Chemical Engineering, The University of Newcastle upon Tyne

3668616

536.2

Edward Arnold

© J. R. Backhurst, J. H. Harker and J. E. Porter 1974
First published 1974
by Edward Arnold (Publishers) Ltd.
41 Bedford Square
London WC1B 3DQ

Reprinted with corrections 1980

ISBN 0 7131 33279 Paper Edition

All Rights Reserved. No part of this publication
may be reproduced, stored in a retrieval system,
or transmitted in any form or by any means,
electronic, mechanical, photocopying, recording
or otherwise, without the prior permission of
Edward Arnold (Publishers) Ltd.

Typeset by Preface Limited, Salisbury, Wilts
and printed in Great Britain by
Unwin Bros. Ltd., The Gresham Press, Old Woking, Surrey

Preface

Chemical engineering is a practical discipline, the essence of which is the solution of problems, whether they be in the design and sizing of equipment, the evaluation of the performance of a particular plant item or in process and plant development. The training of chemical engineers, certainly at the undergraduate level, has two prime objectives; that of teaching a method of attack – a training of the mind to cope with technological problems, and, secondly, an amassing of facts and techniques by which such problems can be handled. The standard texts in the subject lean heavily on the latter objective; the practical application of established techniques, which after all is the life-blood of the chemical engineer, is left to the enterprise of the individual. To take an example, there is little point in grasping fully the fundamentals of drying operations, with all the complexities of diffusion through porous solids, unless one is able to put this into practice in the sizing of a dryer or in the evaluation of drying time, coupled with an understanding of the limitations of the method and the accuracy of the results obtained. The prime aim of this book, therefore, is to illustrate in a fairly simple way, the application of basic chemical engineering techniques and relationships by means of selected worked examples together with a brief summary of the fundamentals involved. The book is not in any way intended to provide an alternative to standard texts but rather to offer a practical guide to the interpretation and application of basic techniques.

The book is divided into three sections; an introduction to the topic, a section on the basic concepts of heat and mass transfer and fluid flow, and a final section dealing with the application of these concepts in the sizing of equipment. The first section considers factors common to all problems in chemical engineering and particular emphasis is placed on methods of attack, and estimation of data and units. In this latter respect, the SI system of units is adopted throughout. Fundamental mass and energy balances are also considered at this stage. The second section provides a brief review of the basic building blocks of chemical engineering – heat and mass transfer. Again, this is not meant as a full treatment of the subject, as this is already available in standard texts, but rather an indication of how fundamental data may be obtained from practical measurements. As processes of heat and mass transfer in fluids are governed by their dynamic behaviour, a chapter on basic fluid mechanics has also been included in this section. Space limitations have dictated that the notes and problems in this chapter be restricted to the definition and development of simple concepts. Experience has shown, however, that adequate attention to the assimilation of fundamental ideas is an essential foundation for their subsequent application to similar and related problems. The final section deals with calculations for the design of equipment in which heat transfer and the important unit operations, distillation, gas absorption and solvent extraction are carried out. A final chapter considers combined heat and mass transfer in humidification and drying operations.

The level of treatment is that of the undergraduate honours course and it is hoped that the book will prove useful to students at all stages of their undergraduate endeavours. For this reason, many of the problems are suited to examination purposes and indeed, several are based on those set in the examinations of the University of Newcastle upon Tyne, whose cooperation in permitting their use in this book is gratefully acknowledged. Whilst written with the needs of the chemical engineer in mind, it is hoped that the book will prove useful also to students in allied disciplines, both in pure and applied science, and also to practising engineers, who may find several sections of value in providing a simple and rapid revision of techniques.

The selection of material has been one based largely on personal choice and yet we have tried to cover a wide field within the one volume. Many of the problems are idealized in that only pure materials are considered for which data are readily available and a unique answer is possible. If we are guilty of omission in this respect, our defence is that our aim is the illustration of techniques rather than the evaluation of complex industrial problems.

No book of this nature can be written in isolation and we gratefully acknowledge the enormous contribution to our knowledge and under-standing made by the standard texts in chemical engineering. Those we have used extensively are listed in a short bibliography at the end of each chapter. We are also grateful to our colleagues at the University of Newcastle upon Tyne, particularly to Professor J. D. Thornton, Mrs Joan

Doherty for careful preparation of the typescript, and also to our publishers for a great deal of efficient work in the final preparation of this text.

Newcastle upon Tyne.   J R Backhurst
                                                         J H Harker
                                                         J E Porter

List of worked examples

List of worked examples

List of worked examples

Contents

Contents

Contents

1. Units and Data

1.1 Introduction

One of the early lessons derived from a training in pure or applied science is an appreciation of the need for a systematic approach to all areas of study. Since problem-solving is one of the fundamental tasks of the engineer, irrespective of his particular discipline, the application of systematic methods to this activity is essential. Faced with a particular problem, how is the engineer to sketch out the possible route, or routes, to a solution? are there any general rules to the game? In this and subsequent chapters, a variety of problems is specified and solved, and it will be apparent that a simple classification can be applied to the problems themselves. Such a classification is an important step towards establishing a policy for problem solution.

However, the engineer is concerned with more than technique; he must produce the correct solution to the problem. In view of the very wide range of materials which the chemical engineer may be required to handle, some guidance on the provision of accurate physical property data is essential and thus it forms a section of this chapter. There is also the need for a consistent system of units. The SI system has been used throughout this book. However, it is apparent that several systems of units will be in use for many years to come, and the engineer must be prepared and able to make the necessary conversions between them.

1.2 Problem solving

In general, problems of heat or mass transfer seem to fall into three main categories:

1

  a. Problems requiring the application of a well established design technique.
  b. Problems which are similar, or analogous, to those for which an analysis has already been developed.
  c. Problems which demand a return to first principles.

In the following chapters, problems which fall into all these categories are solved and the reader should attempt to identify the technique that has been applied in each case.

It should be noted that all problems, at least initially, belong to the third category. Whilst the engineer may use methods of solution appropriate to the first and second categories for convenience, the correct application of a design technique, or recognition of the analogous problem demands a considerable knowledge of underlying principles. Failure to understand and observe the conditions under which 'off the shelf' techniques are valid can lead to incorrect solutions.

1.3 Physical property data

Since all equations that describe rates of heat or mass transfer relate one or more of the physical properties of the solids of fluids involved, then at some stage in the calculation of these quantities such data must be provided. For example, the heat transfer coefficient between a surface and an adjacent fluid may depend upon the density, viscosity, specific heat and thermal conductivity of the fluid in question. On the other hand, if the engineer is concerned with mass transfer between adjacent phases, then other parameters, notably the diffusivities of species undergoing transfer and their equilibrium concentrations in each phase, are extremely important. In these and any other situation, a particular problem cannot be solved until this basic information is available.

1.3.1 Rheological data

The measurement of shear stress and shear rate in a fluid, and the definition and calculation of viscosity are discussed in chapter 3. When the ratio of shear stress to shear rate is constant, that is to say the viscosity of the fluid is independent of shear rate, then the fluid is said to be Newtonian. Viscosity data for a wide range of liquids and gases are available in the standard literature (1–4, 7) or from two other main sources;

(a) organizations that disseminate and tabulate data (5), and
(b) organizations that provide research literature retrieval guides (6).

In addition to these sources of information, there is always the possibility of estimating the required property from the chemical nature (or from any other available data) of the fluid in question. The most notable contribution to the field of physical property estimation is the detailed work of Reid and Sherwood (7) which can usually be used to provide approximate values, at least, of the physical properties of less common fluids. The techniques listed in reference (7) have been incorporated into a comprehensive computer programme (8) for data estimation. This programme can be used to estimate the relevant properties of any chemically well-defined compound or mixture from a little basic information. This is done by searching the library of estimating methods and selecting the most reliable for the information which is available. The programme indicates also the probable error margins associated with an estimated value.

Unfortunately, many process liquids do not exhibit Newtonian behaviour and even elementary engineering calculations require considerable rheological data. In this case, measurement of rheological parameters, say shear stress and shear rate, must be undertaken. These measurements and the subsequent analysis of these data are discussed in chapter 3.

1.3.2 Thermo-physical properties

Two properties are of direct interest under this heading; thermal conductivity and heat capacity. The definition of thermal conductivity and its measurement and calculation for a solid are detailed in chapter 4. Similar measurements in gases or liquids are made difficult by the need to ensure that the 'true' conductivity is measured. This means that elaborate precautions must be taken to ensure that convection effects are reduced to insignificant levels. An excellent survey of the literature on thermal conductivity measurements in liquid can be found in the report of Jamieson and Tudhope (9).

Liquids containing solids (slurries) present a further difficult problem. Except at low particle concentrations, the estimation of the thermal conductivity of a slurry using the conductivities of the liquid and solid phases and the solid concentration is hazardous, and usually this parameter is determined experimentally. For low particle concentrations (<10%) of an inert solid, methods of estimation are available (10, 11).

The heat capacity of solids, single liquids or liquids are usually available from the sources indicated earlier in this section. In dilute suspensions, the contribution of the solid can be ignored. If the solids are inert then a weighted mean value can be used. In almost all other cases, experimentally determined values will be necessary.

1.3.3 Equilibrium data

In mass transfer operations, data are required relating the composition of a component in a liquid mixture to its composition in the vapour above that liquid. Experimental data have been obtained and published for many systems (1, 12) and use should be made of these data wherever possible. However, for the majority of systems, such data are unobtainable and we have to estimate the required equilibrium data.

The basis of all prediction methods for x-y data is the vapour pressure of the pure components of the system; this information is readily available for many compounds (1, 4, 13). Two procedures then follow, depending upon whether the system is ideal or non-ideal. For ideal systems we use Raoult's Law to calculate the required vapour-liquid equilibrium data — a typical example is included in chapter 7. Where the system is known to be non-ideal, activity coefficients and fugacities are introduced into Raoult's Law to allow for the degree of non-ideality. Experimental data for activity coefficients are thinly spread throughout the literature so that after an initial check with Perry (1), we use thermodynamic relationships for the prediction of the missing data. The relevant equations are referred to in chapter 8 where this problem is discussed in greater detail.

The petroleum industry makes use of 'K' factors for equilibrium relationships, where $K = y/x$ and values of K are presented as a function of temperature for many compounds commonly encountered in that industry. Again Perry (1) is a good starting point for such data and further reference may be made to references from that text.

Basic diffusivity data are discussed at some length in chapter 5 and both experimental and empirical correlations for the estimation of the diffusivities of gases and liquids are considered.

We have already mentioned the methods of prediction of physical properties presented in Reid and Sherwood (7) and for an excellent and concise summary of the methods available refer to Holland et al (14) who, incidentally, use SI units throughout.

1.4 Dimensions and units

Before any calculations are carried out, it is most important that all quantities are expressed in a consistent system of units. Unlike the pure scientist, the chemical engineer must be familiar with many systems of units and be able to convert readily quantities from one system to another. This arises from the fact that much basic data, especially relating to physical properties, are evaluated in the laboratory in the centimetre-gramme-second (c.g.s.) system of units, whereas, on the plant scale, quantities have hitherto been expressed in the foot-pound-hour (f.p.h.) system and more recently in the SI system of units. Since chemical engineering has its origin in well-established industries, many forms of empirical unit are also still in use, reflecting early tests for evaluating a particular property, developed before the scientific principles were understood fully. An example of this type of unit is the degree Twaddel ($^\circ$Tw) — an empirical unit of density the scientific unit being g/cm^3, where $^\circ$Tw = 200 (sg − 1) and sg is the specific gravity of the liquid. In engineering work, density is expressed in lb/ft^3; where 1 g/cm^3 = 62.4 lb/ft^3. The general adoption of the SI system of units, in which density is expressed in kg/m^3, will go a long way towards rationalizing units and simplifying the work of the chemical engineer.

1.4.1 Dimensions

An important feature of chemical engineering is the correlation of data in the form of dimensionless groups of properties and other quantities. Perhaps the best known of such a group is specific gravity, being the ratio:

$$\frac{\text{mass of a volume of liquid}}{\text{mass of the same volume of water at a specified temperature}}$$

Since this is a ratio of masses, specific gravity has no dimensions and hence no units. In deciding whether or not a particular group of quantities is dimensionless or not, it is important to be able to reduce quantities to their fundamental dimensions, usually length L and time T and either force F or mass M. Force is used more ofoften in engineering work, though in the SI system of units and in general scientific work, mass is the fundamental dimension. Other quantities may now be expressed in terms of these fundamental dimensions. For example, velocity becomes L/T and acceleration L/T^2. Force is then ML/T^2 and pressure, force per unit area, $ML/T^2L^2 = M/LT^2$. Dynamic viscosity is force per unit area per unit velocity gradient and is therefore $(ML/T^2)/(L^2)(L/TL) = M/LT$ and similarly, kinetic viscosity, which is the ratio of dynamic viscosity to density, is $(M/LT)/(M/L^3) = L^2/T$. Energy, the product of force and distance, is ML^2/T^2 and hence heat, that is thermal energy, has the same dimensions. In practice, however, it is more convenient to express temperature and heat as additional fundamental dimensions. One way is to express heat as $M\theta$, where θ is temperature in K, which, in this definition, would have the dimensions L^2/T^2.

Example 1-1: Show that the Nusselt Number, hd/k, the Reynolds Number, $du\rho/\mu$ and the Prandtl Number, $c_p\mu/k$, are dimensionless by reducing them to the fundamental dimensions M, L, T and θ.

Solution

(i) *Nusselt number*

　　h, the heat transfer coefficient is (thermal energy)/(time)(area) (temperature difference)

$$= M\theta/TL^2\theta = M/L^2T.$$

　　d, the characteristic length $= L$

　　k, thermal conductivity is (thermal energy)/(time)(area)(temperature difference)/(length)

$$= \theta ML/L^2 T\theta = M/LT$$

$$\therefore \ Nu = (M/L^2T)L/(M/LT)$$

$$= (M/L^2T)/(M/L^2T)$$

and is therefore dimensionless.

(ii) *Reynolds Number*

　　d, the characteristic length $= L$

　　u, velocity $= L/T$

　　ρ, density $= M/L^3$

　　μ, dynamic viscosity $= M/LT$

$$\therefore \ Re = L(L/T)(M/L^3)/(M/LT)$$

$$= (M/LT)/(M/LT)$$

and is therefore dimensionless.

(iii) *Prandtl Number*

The ratio, $Nu/Re = (hd\mu)/(kdu\rho) = h\mu/(ku\rho)$

$$= Pr(h/c_p u\rho)$$

Since Nu and Re are dimensionless, Pr is also dimensionless providing the ratio $(h/c_p u\rho)$ has no dimension. This ratio is equivalent to

$$(M/L^2T)/(M\theta/M\theta)(L/T)(M/L^3) = ML^3T/(ML^3T)$$

and hence Pr is also dimensionless.

1.4.2 Systems of units

A system of units is characterized by the basic units of the fundamental dimensions, as outlined in the previous section. Rather than express all quantities in terms of the basic units, however, it is more convenient to give a specific group of the basic units a special name, i.e. a derived unit. In essence, this was done in section 1.4.1 with the group L^2/T^2 (or cm^2/s^2 in the c.g.s. system) which was given the new unit, temperature in K. Force is a good example of the derived unit. In the c.g.s. system, the unit of force is the dyne, which is that force which will give a mass of 1 g an acceleration of 1 cm/s^2. Similarly, in the foot-pound-second system, the unit of force is the poundal, where 1 poundal $= 1 \text{ lb ft/s}^2$. To take another example, 1 poise is the unit of dynamic viscosity in the c.g.s. system ($1 \text{ P} = 1 \text{ g/cm s}$), whilst in British units, no special name has been given to the unit and viscosity is measured in lb/ft h or lb/ft s.

In keeping with recent trends, both in industry and commerce, this text uses the SI (Système International d'Unités) system of units throughout. In this system, the basic units, particularly relevant to chemical engineering are mass – kg, length – m, time – s, temperature – K and electric current – A. The system abounds in derived units and also multiplying prefixes. The most important derived units are

force – the newton, $1 \text{ N} = 1 \text{ kg m/s}^2$

work, energy, heat – the joule, $1 \text{ J} = 1 \text{ Nm} = 1 \text{ kg m}^2/\text{s}^2$

power – the watt, $1 \text{ W} = 1 \text{ J/s} = 1 \text{ kg m}^2/\text{s}^3$

Unlike the British gravitational system of units, where the gravitational constant, g, appears unexpectedly from time to time, SI is a coherent system of units and, for example, the weight of a mass M kg is a force of Mg N where g, the local gravitational acceleration, is in m/s^2. Many groups of units are without special names and these are usually expressed in terms of the derived unit rather than the basic units. Thus, the unit of pressure is N/m^2, dynamic viscosity, $N\ s/m^2$, kinematic viscosity, m^2/s, thermal conductivity, $W/m\ K$ and a heat transfer coefficient is expressed in $W/m^2 K$.

In many cases, it is convenient to make use of multiplying prefixes to give reasonably sized values of quantities. The most common of these are

$$10^{-6}: \text{micro-}, \mu \qquad 10^6: \text{mega-}, M$$

$$10^{-3}: \text{milli-}, m \qquad 10^3: \text{kilo-}, k$$

$$10^{-2}: \text{centi-}, c$$

only one prefix being allocated to a given unit at any one time. Thus, instead of writing 1 year = $(365 \times 24 \times 3600) = 31\ 536\ 000$ s, it is more convenient to express this as $31 \cdot 5$ Ms. Similarly, 1 atmosphere becomes $101 \cdot 3$ kN/m^2 rather than $101\ 325$ N/m^2.

1.4.3 Conversion of units

Before commencing any calculation, it is vital that all quantities involved are expressed in a consistent system of units. With the SI system, it is equally important that quantities that are unity in other systems are not omitted, as is often the custom with c.g.s. units. For example, the heat capacity of water, which is 1 cal/g $^\circ$C = 1 Btu/lb $^\circ$F becomes $4 \cdot 187$ kJ/kg K. Similarly, the density and viscosity of water are no longer unity, but 10^3 kg/m^3 and 10^{-3} N s/m^2 respectively. As a bonus however, the approximate density and heat capacity of air are 1 kg/m^3 and 1 kJ/kg K. Table 1-1 includes some of the more important conversion factors, and their use is illustrated in the following examples.

Example 1-2: Water is flowing through a pipe of 1 in inside diameter at a velocity of 3 ft/s. Calculate the Reynolds Number if the mean temperature is 60 $^\circ$F, and check the result by means of the same calculation in SI units.

Solution
i) *British units*

$$d = 1 \text{ in} = 1/12 = 8 \cdot 33 \times 10^{-2} \text{ ft}$$
$$u = 3 \text{ ft/s} \equiv 3 \times 3600 = 1 \cdot 08 \times 10^4 \text{ ft/h}$$

Table 1-1   Factors for Conversion to SI Units

mass		**pressure** (continued)	
1 lb	: 0·454 kg	1 in water	: 249 N/m²
1 ton	: 1016 kg	1 in Hg	: 3·39 kN/m²
		1 mm Hg	: 133 N/m²
length			
1 in	: 25·4 mm	**power**	
1 ft	: 0·305 m	1 hp	: 745 W
1 mile	: 1·609 km	1 Btu/h	: 0·293 W
time		**viscosity**	
1 min	: 60 s	1 P	: 0·1 N s/m²
1 hr	: 3·6 ks	1 lb/ft h	: 0·414 mN s/m²
1 day	: 86·4 ks	1 stoke	: 10⁻⁴ m²/s
1 year	: 31·5 Ms	1 ft²/h	: 0·258 cm²/s
area		**mass flow**	
1 in²	: 645·2 mm²	1 lb/h	: 0·126 g/s
1 ft²	: 0·093 m²	1 ton/h	: 0·282 kg/s
		1 lb/ft²h	: 1·356 g/m² s
volume			
1 in³	: 16 387·1 mm³	**thermal**	
1 ft³	: 0·0283 m³	1 Btu/h ft²	: 3·155 W/m²
1 UK gal	: 4546 cm³	1 Btu/h ft² °F	: 5·678 W/m² K
1 US gal	: 3786 cm³	1 Btu/lb	: 2·326 kJ/kg
		1 Btu/lb °F	: 4·187 kJ/kg K
force		1 Btu/h ft °F	: 1·731 W/m K
1 pdl	: 0·138 N		
1 lbf	: 4·45 N	**energy**	
1 dyne	: 10⁻⁵ N	1 kWh	: 3·6 MJ
		1 therm	: 105·5 MJ
energy			
1 ft lbf	: 1·36 J	**calorific value**	
1 cal	: 4·187 J	1 Btu/ft³	: 37·26 kJ/m³
1 erg	: 10⁻⁷ J	1 Btu/lb	: 2·326 kJ/kg
1 Btu	: 1·055 kJ		
		density	
pressure		1 lb/ft³	: 16·02 kg/m³
1 lbf/in²	: 6·895 kN/m²		
1 atmos	: 101·3 kN/m²		
1 bar	: 10⁵ N/m²		
1 ft water	: 2·99 kN/m²		

From tables, the properties of water at $60°F$ are

$$\rho = 0.998 \text{ g/cm}^3 \equiv 0.998 \times 62.43 = 62.30 \text{ lb/ft}^3$$

$$\mu = 1.1 \text{ cP} = 1.1 \times 2.42 = 2.66 \text{ lb/ft h}$$

$$\therefore Re = du\rho/\mu$$

$$= 8.33 \times 10^{-2} \times 1.08 \times 10^4 \times 6.23 \times 10/2.66$$

$$= 2.11 \times 10^4$$

ii) *SI units*

$$d = 8.33 \times 10^{-2} \text{ ft} \equiv (8.33 \times 10^{-2} \times 0.305) = 2.54 \times 10^{-2} \text{ m}$$

$$u = 3 \text{ ft/s} \equiv (3 \times 0.305) = 0.915 \text{ m/s}$$

$$\rho = 0.998 \text{ g/cm}^3 \equiv 0.998 \times 10^3 = 9.98 \times 10^2 \text{ kg/m}^3$$

$$\mu = 1.1 \text{ cP} = 1.1 \times 10^{-2} \text{ P} \equiv (1.1 \times 10^{-2} \times 0.1) = 1.1 \times 10^{-3} \text{ N s/m}^2$$

$$\therefore Re = 2.54 \times 10^{-2} \times 0.915 \times 9.98 \times 10^2/(1.1 \times 10^{-3})$$

$$= 2.11 \times 10^4$$

(A check on the units of Re gives $m(m/s)(kg/m^3)/(N \text{ s/m}^2)$

$$= (kg \text{ m/s}^2)/N$$

$$= (kg \text{ m/s}^2)/(kg \text{ m/s}^2) \text{ i.e. dimensionless.})$$

Example 1-3: For convective heat transfer to water flowing in a tube, the film coefficient is given by

$$h = 150(1 + 0.011T)u^{0.8}/d^{0.2} \quad \text{Btu/h ft}^2 \, °F$$

where T is the mean water temperature ($°F$)
u is the water velocity (ft/s)
d is the inside diameter of the tube (in)

What is the equivalent equation in SI units? The result should be checked using the data given in Example 1-2.

Solution   If $\underline{T}$ is the temperature (K), $\underline{u}$ the velocity (m/s) and $\underline{d}$ the pipe diameter (m) in SI units, then

$$T = 1.8(\underline{T} - 273) + 32 = 1.8\underline{T} - 460 \qquad °F$$

$$u = \underline{u}/0.305 = 3.28\underline{u} \qquad \text{ft/s}$$

$$d = \underline{d} \times 10^3 \text{ mm} \equiv (\underline{d} \times 10^3/25\cdot4) = 39\cdot4\underline{d} \qquad \text{in}$$

$$\therefore h = 150(1 + 0\cdot011(1\cdot8\underline{T} - 460))(3\cdot28\,\underline{u})^{0\cdot8}/(39\cdot4\underline{d})^{0\cdot2}$$

$$= (3\cdot67\underline{T} - 752)\underline{u}^{0\cdot8}/\underline{d}^{0\cdot2} \qquad\qquad \text{Btu/h ft}^2\,^\circ\text{F}$$

$$\therefore h = 5\cdot678\ (3\cdot67\,T - 752)u^{0\cdot8}/d^{0\cdot2}$$

$$= (20\cdot9\underline{T} - 4280)\underline{u}^{0.8}/\underline{d}^{0\cdot2} \qquad\qquad \text{W/m}^2\text{ K.}$$

Check Calculation
i) *British units*

$$T = 60\,^\circ\text{F}, u = \text{ft/s}, d = 1 \text{ in}$$

$$\therefore h = 150(1 + 0\cdot011 \times 60)3^{0\cdot8}/1^{0\cdot2}$$

$$= 600 \text{ Btu/h ft}^2\,^\circ\text{F}$$

$$\underline{h} = 600 \times 5\cdot678$$

$$= 3407 \text{ W/m}^2\text{ K or } 3\cdot41 \text{ kW/m}^2\text{ K.}$$

ii) *SI units*

$$\underline{T} = ((60 - 32)/1\cdot8) + 273 = 288 \text{ K}$$

$$\underline{u} = (3 \times 0\cdot305) = 0\cdot915 \text{ m/s}$$

$$\underline{d} = (1 \times 25\cdot4/10^3) = 0\cdot0254 \text{ m}$$

$$\therefore \underline{h} = (20\cdot9 \times 288 - 4280)0\cdot915^{0\cdot8}/0.0254^{0\cdot2}$$

$$= 3400 \text{ W/m}^2\text{ K or } 3\cdot40 \text{ kW/m}^2\text{ K}$$

References

1 PERRY, J. H. (Ed). *Chemical Engineer's Handbook*. 4th Edn. McGraw-Hill Publishing Co. New York (1963).
2 ANON. *Thermophysical Properties of Matter*. Plenum Data Corporation New York (1969–1971).
3 AMERICAN PETROLEUM INSTITUTE A.P.I. Project 44. *Selected Properties of hydrocarbons and related compounds* (1966)
4 MAXWELL, J. B. *Data Book on Hydrocarbons*. D. Van Nostrand Co. Inc. New Jersey (1957).
5 i) ENGINEERING SCIENCES DATA UNIT. Institution of Chemical Engineers Belgrave Sq. London
   ii) THERMOPHYSICAL PROPERTIES RESEARCH CENTRE. Purdue University West Lafayette Indianna U.S.A.

iii) HEAT TRANSFER AND FLUIDS SERVICE. U.K.A.E.A., A.E.R.E. Harwell Didcot Berks.

6 TOULOUKIAN, Y. S. (Ed) *Thermophysical Properties Research Literature Retrieval Guide.* 2nd Edn. McDonald & Co. Ltd. London (1968).

7 REID, R. C. & SHERWOOD, T. K. *The Properties of Gases and Liquids.* McGraw-Hill Book Co. New York (1969).

8 APPES N. E. L. *Data Estimation Services.* National Engineering Laboratory East Kilbride Glasgow.

9 JAMIESON, D. T. & TUDHOPE, J. S. *The thermal conductivity of liquids* NEL Report No. 137 (1964) National Engineering Laboratory. East Kilbride Glasgow.

10 ORR, JNR, C. & DALLA VALLE, J. M. Heat transfer properties of liquid-solid suspensions *Chem. Eng. Procy.*, 1954, **50**, No. 9, 29.

11 THOMAS, D. G. Non-Newtonian suspensions *Ind. Engng. Chem. Ind. (int) Edn.*, 1965, **55**, No. 112, 18.

12 HALA, E. et al, *Vapour liquid equilibrium.* Pergamon Press Oxford (1958).

13 GALLANT, R. W. *Physical Properties of Hydrocarbons.* Vols. I and II, Gulf Publishing Co. Houston Texas (1970).

14 HOLLAND, F. A., MOORES, R. M., WATSON, F. A. WILKINSON, J. K. *Heat Transfer.* Heinneman Educational Books London (1971).

Further reading

BRIDGMAN, P. W. *Dimensionless Analysis.* Yale University Press (1931).

BRITISH STANDARDS INSTITUTION *The Use of SI Units.* Pubn. PD .5686 London (1967).

ANDERTON, P. & BIGG, P. H. *Changing to the Metric System.* H.M.S.O. London (1965).

EDE, A. J. The International System of Units. *Int. J. Heat Mass Transfer*, **9**, 837, (1966).

MULLIN, J. W. SI units in chemical engineering *Chem. Engr.*, **211**, CE176, (1967).

LEES, F. P. An SI unit conversion table for chemical engineering, *Chem. Engr.*, **212**, CE341, (1968).

Problem 1-1: For the condensation of vapours on the outside of horizontal tubes, the mean film coefficient is given by

$$h_m = 0 \cdot 72((k^3 \rho^2 g\lambda)/(d_0 \mu \Delta T_f))^{0 \cdot 25} \qquad \text{Btu/h ft}^2 \ ^\circ\text{F}$$

where λ (Btu/lb) is the latent heat of the condensate and ΔT_f ($^\circ$F) is the temperature difference across the condensate film, all other symbols having their usual significance. Evaluate the dimensions of the right hand side of this equation and hence obtain an expression for h_m in W/m^{2}K, all other quantities being expressed in SI units.

$$((M^4/L^8 T^4)^{0 \cdot 25} ; \quad h_m = 0 \cdot 72((k^3 \rho^2 g\lambda)/(d_0 \mu \Delta T_f))^{0 \cdot 25})$$

Problem 1-2: The radiative heat transfer between two planes of emissivities ϵ_1 and ϵ_2 is given by

$$Q = \sigma A(T_1^4 - T_2^4)/((1/\epsilon_1) + (1/\epsilon_2) - 1) \qquad \text{Btu/h}$$

where A (ft^2) is the area of the planes at temperatures T_1 and T_2 ($^\circ$R) and σ is the Stefan–Boltzmann Constant ($0 \cdot 173 \times 10^{-8}$ Btu/h ft^2 $^\circ$R^4) Calculate the Stefan–Boltzmann Constant in SI units and check the result by evaluating the heat transfer for the following conditions

$$T_1 = 200 \ ^\circ\text{F}, \ T_2 = 40 \ ^\circ\text{F}$$

$$A = 224 \ \text{ft}^2$$

$$\epsilon_1 = 0.80, \quad \epsilon_2 = 0 \cdot 91.$$

$$(5 \cdot 67 \times 10^{-8} \ \text{W/m}^2 \ \text{K}^4 ; 36 \ 510 \ \text{Btu/h} \equiv 10 \cdot 6 \ \text{kW})$$

2. Mass and Heat Balances

2.1 Basic concepts of mass balance

The law of conservation of mass states that mass is neither created nor destroyed in any process. Thus for any plant or process

$$(\text{mass in}) = (\text{mass out}) + (\text{accumulation within the system})$$

This statement applies to all types of process involving both physical and chemical operations; and also to the individual components handled within a process, as well as the total mass. Although the statement refers to mass, if all quantities are related to the same fixed period of time, then the principle may be applied also to flowrates of material. If the accumulation or the rate of accumulation within the system is zero, then operation is taking place at *steady-state* and the mass of material within the system is irrelevant in mass-balance calculations. Where there is a finite rate of accumulation (or depletion) within a system, then operation is at *unsteady-state* and calculations are somewhat more specialized. In this latter case, we cannot ignore the physical volume of the plant nor the hold-up of material in the system. In addition, we cannot assume that physical and chemical changes take place instantaneously — an important feature of steady-state calculations.

In mass-balance calculations, there are several points to note. In addition to making an overall mass balance, it is usually necessary to carry out component balances in order to specify the system completely. Where physical operations alone are involved, the number of components is the number of distinct chemical species that exist in all parts of the plant or unit. In operations involving chemical reaction, some if not all the species are related through chemical equations, and the number of components is the number of species minus the number of independent chemical equations. For example, where carbon is burned in air to form a mixture of carbon monoxide and carbon dioxide, five chemical species are involved $(C, O_2, N_2, CO$ and $CO_2)$ and yet these are linked by the two chemical equations

$$C + \tfrac{1}{2}O_2 = CO$$
$$C + O_2 = CO_2$$

There are therefore $(5 - 2) = 3$ components, C, O and N, and balances of these components should be made in addition to the overall balance.

It is important to realize that, where a chemical reaction is involved, it is not usually possible to make a balance in terms of the kilogramme mole (kmol) (1 kmol = molecular weight of an element or compound in kg), since in many reactions the stoichiometric ratio of products to reactants is not unity. The 'kmol' does provide a useful basis, however, in assessing quantitatively the products formed in a chemical reaction and also in making a balance where physical processing alone is involved. This is especially true in systems where gaseous components are being handled. In all chemical reactions, the number of kmol of a selected element in the reactants is always equal to the kmol of that element in the products. This forms an obvious and most useful form of mass balance. Similarly, it is equally convenient to make balances of the kmol of suitable radicals in the products and reactants. An obvious example is SO_3, in which case H_2SO_4 would be written as $H_2O.SO_3$. Other suitable radicals are P_2O_5, N_2O_5, K_2O, etc.

Where one reactant forms two or more products, it is a much sounder procedure to use the simplest possible chemical equations, rather than one complex overall equation representing the entire reaction. In this way, confusion as to the exact ratio of components in a feed stream is avoided.

2.2 Mass balance at steady-state

2.2.1 Basic problems

Essentially, there are two types of basic problem. In the first, the compositions of all streams are known together with the mass (or flowrate) of one or more process streams. The mass (or flowrate) is then obtained by solving a series of simultaneous equations. In the second type of problem, not all product analyses are known and these may be obtained by a progressive calculation from the feed and other input conditions.

Example 2-1: In order to separate pure alcohol from alcohol-water mixtures, it is necessary to add a third component, such as benzene, to reduce the volatility of the alcohol and hence produce pure alcohol as the bottom product. In such an operation, a benzene-free feed containing 88·0% alcohol and 12·0% water by mass gives an overhead product containing 17·5% alcohol, 7·9% water and 74·6% benzene by mass. What volume of benzene must be fed to the column per unit time in order to produce 1250 cm³/s pure alcohol as the bottom product and what percentage of the alcohol in the feed is obtained as absolute alcohol? (density of benzene = 870 kg/m³, density of alcohol = 785 kg/m³).

Solution. The various flows are shown in Fig. 2-1, where F kg/s is the alcohol-water feed, B kg/s is the flow of the added benzene and, conventionally, D kg/s and W kg/s are the flowrates of distillate and bottom product respectively.

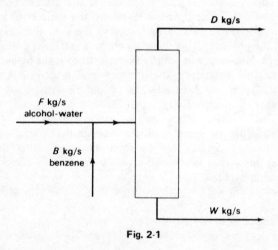

Fig. 2-1

The bottom product = 1250 cm³/s = 0·00125 m³/s
Hence

$$W = (785 \times 0·00125) = 0·981 \text{ kg/s}$$

Then

overall balance $F + B = D + 0·981$

alcohol balance $(88·0F/100) = 17·5D/100) + 0·981$,
              or $F = 0·20D + 1·12$

water balance $(12 \cdot 0F/100) = 7 \cdot 9D/100)$ or $F = 0 \cdot 66D$

benzene balance $B = (74 \cdot 6D/100)$ or $B = 0 \cdot 75D$

Substituting in the overall balance

$$0 \cdot 66D + 0 \cdot 75D = D + 0 \cdot 981$$
$$D = 2 \cdot 40 \text{ kg/s}$$

and $B = (0 \cdot 75 \times 2 \cdot 40) = 1 \cdot 8 \text{ kg/s}$

or $(1 \cdot 8/870) = 2 \cdot 07 \times 10^{-3} \text{ m}^3/\text{s}$

$$F = (0 \cdot 66 \times 2 \cdot 40) = 1 \cdot 58 \text{ kg/s}$$

Flow of alcohol in the feed $= (1 \cdot 58 \times 88/100) = 1 \cdot 40 \text{ kg/s}$, and the percentage alcohol recovered in the bottom product

$$= (0 \cdot 981 \times 100/1 \cdot 44)$$
$$= 70 \cdot 5\% \text{ by mass.}$$

Example 2-2: It is desired to produce 100 kg of a solution of ammonia in water containing 10% by mass of ammonia. This will be achieved by stripping ammonia from an ammonia-air mixture by water in a counter-current absorption tower. If the inlet gas stream contains 20% ammonia by mass and the outlet gas 2% ammonia by mass what mass of ammonia-air mixture should be fed to the column?

Solution A simplified flow diagram is included as Fig. 2-2. In 100 kg liquid product, mass of ammonia $= (100 \times 10/100) = 10 \text{ kg}$ and mass of

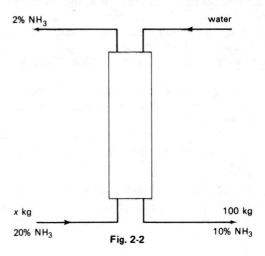

2% NH$_3$ water

x kg   100 kg

20% NH$_3$ **Fig. 2-2**   10% NH$_3$

water = $(100 - 10) = 90$ kg, this being equal to the mass of water fed to the column. If x kg ammonia-air mixture is fed to the column, then mass of air in the inlet gas stream = $(100 - 20)x/100 = 0\cdot8x$ kg, this is equal to 98% of the exit gas stream and hence,

mass of exit gas stream = $(100 \times 0\cdot8x/98)$

$$= 0\cdot817x \text{ kg.}$$

mass of ammonia in exit gas stream = $(0\cdot817x \times 2/100)$

$$= 0\cdot0163x \text{ kg}$$

Thus, making an ammonia balance,

$$(20x/100) = 10 + 0\cdot0163x$$

$$x = 10/(0\cdot20 - 0\cdot0163)$$

$$= 54\cdot5 \text{ kg}$$

(NB. A common error is to assume that $(20 - 2) = 18\%$ of the inlet gas stream is absorbed by the water. The above calculation shows that the true value is $(20 - 1\cdot63) = 18\cdot37\%$.)

2.2.2 Mass balance with chemical reaction

The previous examples involve only physical operations in that no chemical reaction takes place. Calculations involving combustion processes provide a good example of mass balance with reaction and also illustrate two points; firstly, the use of the 'kmol' of each element as a means of balance, and secondly, linking the various input and output streams by means of an inert component, usually nitrogen. (In fact, the air was used in this way in Example 2-2.)

There are basically two types of combustion calculation:

(a) the calculation of the flue gas analysis and the mass of flue gas for various amounts of excess air, given the composition of the fuel, and

(b) the calculation of the amount of air used, and, in simple cases, the composition of the fuel, given the flue gas analysis.

As with all steady-state, mass balances involving chemical reaction, it is assumed that the reactions take place instantaneously and that they are 100 per cent complete.

Example 2-3:   A furnace is fired with coal of the following composition by weight: C 75·2%, H 4·8%, N 1·3%, S 1·0%, O 8·3%, moisture 4·8%, ash 4·6%. Calculate the theoretical air requirements for the combustion of 1 kg of the fuel and the volumetric analysis of the dry flue gas when the fuel is burned with 50% excess air. What is the mass of flue gas produced?

Solution   The combustion equations in this case are

$$C + O_2 \quad = CO_2$$

$$H_2 + \tfrac{1}{2}O_2 = H_2O$$

$$S + O_2 \quad = SO_2$$

Thus, 12 kg carbon combine with 32 kg oxygen to form 44 kg carbon dioxide. It is much more convenient, however, to calculate the combustion products and the oxygen requirements on the basis of the 'kmol'. That is, kmol carbon combines with 1 kmol oxygen to form 1 kmol of carbon dioxide which occupies 22·4 m^3 at 101·3 kN/m^2 and 273 K. Similarly, 1 kmol of hydrogen reacts with 0·5 kmol oxygen to form 1 kmol water, which may be assumed to condense out. The calculation of theoretical air requirements is best made in tabular form

basis: 100 kg coal as fired

component	% by weight	÷ by	kmol	kmol oxygen required
C	75·2	12	6·27	6·27
H	4·8	2	2·40	1·20
N	1·3	28	0·05	–
S	1·0	32	0·03	0·03
O	8·3	32	0·26	−0·26
moisture	4·8	18	0·27	–
ash	4·6	–	–	–

total O = 7·24 kmol/100 kg coal.

Taking the composition of air as 21% oxygen by volume,

the theoretical air requirements = (7·24 x 100/21) = 34·48 kmol/100 kg coal

volume of air required = (34·48 x 22·4/100) = 7·72 m^3/kg coal at 101·3 kN/m^2 and 273 K

The 'mean molecular weight' of air is approximately

$$(0.21 \times 32) + (0.79 \times 28) = 28.9 \text{ kg/kmol}$$

and hence the mass of theoretical air is

$$(34.48 \times 28.9/100) = 9.97 \text{ kg/kg coal}$$

Now, excess air is

$$100 \text{ (actual } - \text{ theoretical)/theoretical,}$$

therefore, actual air is

$$(34.48 \times 150/100) = 51.7 \text{ kmol/100 kg coal.}$$

This contains

$$(51.7 \times 21/100) = 10.85 \text{ kmol oxygen}$$

and

$$(51.70 - 10.85) = 40.85 \text{ kmol nitrogen.}$$

Oxygen required for combustion is 7.24 kmol, therefore excess oxygen in flue gas is

$$(10.85 - 7.24) = 3.61 \text{ kmol.}$$

The nitrogen in flue gas is

$$40.85 \text{ (from air)} + 0.05 \text{ (from coal)} = 40.90 \text{ kmol.}$$

The combustion products are

$$CO_2 = 6.27 \text{ kmol, } SO_2 = 0.03 \text{ kmol.}$$

Total amount of dry flue gas is therefore

$$(3.61 + 40.90 + 6.27 + 0.03) = 50.81 \text{ kmol.}$$

$$CO_2 \text{ in dry flue gas} = (6.27 \times 100/50.81) = 12.35\%$$
$$SO_2 \text{ in dry flue gas} = (0.03 \times 100/50.81) = 0.06\%$$
$$O_2 \text{ in dry flue gas} = (3.61 \times 100/50.81) = 7.10\%$$
$$N_2 \text{ in dry flue gas} = (40.90 \times 100/50.81) = 80.49\%$$

(This is the analysis that would result from an Orsat analysis for example, as in this test, the water condenses out and the composition is on the dry basis. It should be noted that mol % is equivalent to volume %, as the molecular weight of all gases in kg, occupies the same volume (22.4 m^3 at 101.3 kN/m^2 and 273 K.)

The mass of flue gas is calculated as follows

$$\text{mass of } CO_2 = (6 \cdot 27 \times 44) = 276 \text{ kg}$$
$$\text{mass of } SO_2 = (0 \cdot 03 \times 64) = 2 \text{ kg}$$
$$\text{mass of } O_2 = (3 \cdot 61 \times 32) = 116 \text{ kg}$$
$$\text{mass of } N_2 = (40 \cdot 90 \times 28) = \underline{1145} \text{ kg}$$
$$1539 \text{ kg}$$

In addition, the wet flue gas contains $2 \cdot 40$ kmol water from combustion of the hydrogen and $0 \cdot 26$ kmol water from the moisture in the coal – a total of $2 \cdot 66$ kmol or $(2 \cdot 66 \times 18) = 48$ kg. Thus the total mass of wet flue gas is

$$(1539 + 48) = 1587 \text{ kg} \quad \text{or} \quad 15 \cdot 9 \text{ kg/kg coal}$$

(This value could have been obtained rather more easily by an overall balance:

$$\text{mass of air used} = 9 \cdot 97 \times 150/100 = 14 \cdot 9 \text{ kg}$$
$$\therefore \text{ mass of flue gas } = \text{mass of air} + \text{mass of coal}$$
$$= 14 \cdot 9 + 1 \cdot 0$$
$$= 15 \cdot 9 \text{ kg/kg}$$

This calculation is often more complicated, however, especially where combustion is incomplete, due, for example, to carbon being lost in the ash or in the formation of carbon monoxide.)

Example 2-4 A boiler is fired with a natural gas containing only methane and nitrogen. The analysis of the dry flue gas is: CO_2 $8 \cdot 05\%$, O_2 $6 \cdot 42\%$, N_2 $85 \cdot 53\%$.

Calculate the composition of the fuel and the amount of excess air used in the combustion.

Solution

Basis 100 kmol dry flue gas. Assume the natural gas contains x kmol methane and y kmol nitrogen and that a kmol of air is supplied per 100 kmol dry flue gas.

C balance: 100 kmol flue gas contains $8 \cdot 05$ kmol CO_2 (1 kmol CO_2 contains 1 kmol C), therefore carbon in dry flue gas is $8 \cdot 05$ kmol. In the fuel, x kmol CH_4 each contain 1 kmol C, therefore carbon in fuel is x kmol and hence, $x = 8 \cdot 05$ kmol.

O_2 balance: 100 kmol dry flue gas contains

$$8 \cdot 05 \text{ (in } CO_2) + 6 \cdot 42 \text{ (in } O_2) = 14 \cdot 47 \text{ kmol } O_2$$

21

8·05 kmol CH_4 produce 16·10 kmol water according to the equation

$$CH_4 + 2O_2 = CO_2 + 2H_2O$$

1 kmol water contains 0·5 kmol O_2 and hence oxygen in water formed is 8·05 kmol. Then

(oxygen in air) = (oxygen in dry flue gas) + (oxygen in water)

$$(21a/100) = 14·47 + 8·05$$

$$\therefore a = 107·2 \text{ kmol/100 kmol dry flue gas.}$$

N_2 balance: (N_2 in fuel) + (N_2 in air) = (N_2 in flue gas)

$$y + (79 \times 107·2/100) = 85·53$$

$$\therefore y = 0·84 \text{ kmol.}$$

Thus 0·84 kmol nitrogen is associated with 8·05 kmol methane in the fuel and the analysis is

$$\% \ CH_4 = (8·05 \times 100/(8·05 + 0·84)) = 90·6\%$$

$$\% \ N_2 = (100 - 90·6) \qquad = 9·4\%$$

Excess air: 1 kmol CH_4 requires 2 kmol O_2 for complete combustion, therefore theoretical oxygen is

$$(2 \times 8·05) = 16·10 \text{ kmol/100 kmol dry flue gas}$$

actual oxygen supplied is

$$(21 \times 107·2/100) = 22·57 \text{ kmol}$$

Therefore, excess air is

$$((22·51 - 16·10) \times 100/16·1) = 40\%$$

(This illustrates the use of actual and theoretical oxygen in computing the percent excess air. This is perfectly valid because of the constant composition of air.)

2.2.3 Recycle and by-pass-streams

Streams are recycled in both physical and chemical processes for a variety of reasons, and by-passing of a particular processing step by a proportion of a stream is also fairly common. A chemical reactor in which unreacted feed is separated from the product stream and returned with the feed is an obvious example. The basic principles outlined in section 2.1 apply and the secret of solving problems lies in the correct choice of balance positions. As the overall balance of the entire process involves the net output and input, it is probably the most important from a production standpoint. A balance which excludes the recycle is of equal value, however, especially in sizing of the equipment involved, as the recycle stream flowrate can then be deduced by difference from the overall balance. Further balances around the points of recycle (or by-pass) offtake and re-entry enable the complete analysis of the process to be made.

Example 2-5:   The feed gas to an ammonia plant contains 25% nitrogen and 75% hydrogen, the stoichiometric proportions. 12% of the reactants is converted to ammonia, which is condensed out and the unreacted gas is recycled to the reactor. In order to avoid the build up of argon in the system, a fraction of the recycle gas is bled off. If the volumetric composition of air is N_2 78·03%, O_2 20·99%, A 0·98%, calculate the fraction of the recycle gas which must be removed in order that the maximum concentration of argon entering the reactor is 0·4%.

Solution:
   Balance around the reactor (Fig. 2-3). Assume 100 kmol/s of argon-free nitrogen-hydrogen mixture, that is 25 kmol N_2 and 75 kmol H_2, are fed to the reactor. Thus, if the mixture contains x kmol A, then

$$100\,x/(100 + x) = 0·4$$
$$\therefore \quad x = 0·412 \text{ kmol/s}$$

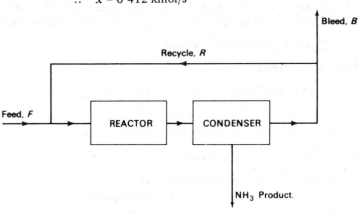

Fig. 2-3

23

Thus the feed to the reactor is 25 kmol N_2, 75 kmol H_2, and 0·412 kmol A. The reactants are in the stoichiometric ratio according to the equation

$$N_2 + 3H_2 = 2NH_3$$

$(25 \times 12/100) = 3·0$ kmol N_2 take part in the reaction together with $(3·0 \times 3) = 9·0$ kmol H_2 forming $(2 \times 3·0) = 6·0$ kmol NH_3. Therefore the outlet stream contains

$N_2 = (25·0 - 3·0) = 22·0$ kmol

$H_2 = (75·0 - 9·0) = 66·0$ kmol

$NH_3 = 6·0$ kmol

$A = 0·412$ kmol (unreacted).

Balance around the condenser

The above stream is fed to the condenser in which the ammonia is completely removed. The gas stream leaving the condenser therefore contains 22·0 kmol N_2, 66·0 kmol H_2 and 0·412 kmol A — a total of 88·412 kmol. The concentration of A in the recycle (and the bleed stream) is

$$(0·412 \times 100/88·412) = 0·466\%$$

Balance around the recycle and feed interchange

Let recycle flow = R kmol/s and feed rate = F kmol/s, then overall balance gives

$$F + R = 100·412$$

In 100 kmol air, 78·03 kmol N_2 are associated with 0·98 kmol A. The feed stream contains 25% N_2 and this is therefore associated with

$$(25 \times 0·98/78·03) = 0·314 \text{ kmol A.}$$

Hence, % A in feed is

$$0·314 \times 100/(100 + 0·314) = 0·313\%$$

Argon balance

$$(0·313 F/100) + (0·466 R/100) = 0·412$$

Substituting for F from the overall balance,

$$0·313(100·412 - R) + 0·466 R = 41·2$$

$$\therefore \quad R = 63·9 \text{ kmol/s}$$

$$\text{and } F = 36·5 \text{ kmol/s}$$

Balance around the bleed and recycle interchange
Let bleed flow = B kmol/s, then total amount of gas leaving condenser
is 88·41 kmol/s. Therefore

$$88\cdot41 = B + R$$
$$B = 24\cdot5 \text{ kmol/s}$$

Hence the fraction of the recycle which is bled off is

$$(24\cdot5 \times 100/63\cdot9) = 38\cdot4\%$$

2.2.4 Problems involving several units

The previous examples deal essentially with the balance around a single
piece of equipment. Practical problems usually involve a number of
plant items, though the aim of the balance — the complete specification
of all streams — remains the same. The data available and the way in
which they are expressed determines the most effective way to achieve
this aim. Two general methods may be considered. Where the available
data relate to one end of the plant, the order in which balances are
made is unimportant. Where information on selected inlet and outlet
streams is given, it is usually necessary to solve for the unknowns in
part of the plant before balances may be made across other units.

Example 2-6: In the U.K. Gas Council's Autothermic Process, a light
distillate is gasified at 2000 kN/m² in two stages:

  i) reaction with steam at 750 K,
  ii) addition of air and second stage reaction at 950 K.

The carbon monoxide in the second stage product is then converted
to carbon dioxide which is removed in a final stage during which the gas
is dried. The gas analyses (per cent by volume) after various stages are
as follows:

%	ex-first stage reaction with steam	ex-second stage reaction with air	after CO conversion	after CO₂ removal & drying
CO₂	13.9	11.8	16.1	1.0
CO	1.3	6.0	1.8	2.9
H₂	12.9	23.7	27.8	44.5
CH₄	36.2	22.0	22.0	35.1
N₂	0	10.3	10.3	10.5
H₂O	35.7	26.2	22.0	0

If the feed to the plant contains 84·8% C and 15·2% H by mass, calculate:

    (a) the ratio steam: feed in the first stage reaction,
    (b) the air supplied in the second stage/kg feed, and
    (c) the final volume of dry gas/kg feed.

Solution   A simplified flow diagram is shown in Fig. 2-4.

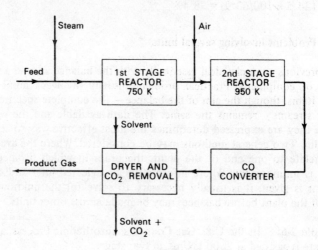

Fig. 2-4

Basis: 100 kg feed
(a) Mass of C in 100 kg feed is 84·8 kg

$$= (84·8/12) = 7·07 \text{ kmol.}$$

In 100 kmol gas leaving the first stage reactor

$$C = 13·9 \text{ (in } CO_2) + 1·3 \text{ (in CO)} + 36·2 \text{ (in } CH_4) = 51·4 \text{ kmol.}$$

C balance: amount of gas leaving the first stage reactor is

$$(7·07 \times 100/57·4) = 13·8 \text{ kmol/100 kg feed.}$$

In 100 kmol gas leaving the 1st stage reactor, H_2 is

$$12·9 \text{ (in } H_2) + 72·4 \text{ (in } CH_4) + 35·7 \text{ (in } H_2O) = 121·0 \text{ kmol.}$$

In 13·8 kmol gas (i.e. equivalent to 100 kg feed), H_2 is

$$13·8 \times 121·0/100 = 16·7 \text{ kmol } H_2/100 \text{ kg feed.}$$

26

H_2 balance:

$$(H_2 \text{ in feed}) + (H_2 \text{ in steam}) = (H_2 \text{ in product})$$
$$(15 \cdot 2/2) + x = 16 \cdot 7$$
$$x = 9 \cdot 1 \text{ kmol } H_2$$

As 1 kmol steam contains 1 kmol H_2, amount of steam supplied is $9 \cdot 1$ kmol

$$= (9 \cdot 1 \times 18) = 164 \text{ kg steam}/100 \text{ kg feed}$$

Therefore, steam:feed ratio is $1 \cdot 64$ kg/kg

(b) In 100 kmol gas leaving second stage reactor, C is

$$11 \cdot 8 \text{ (in } CO_2) + 6 \cdot 0 \text{ (in CO)} + 22 \cdot 0 \text{ (in } CH_4) = 39 \cdot 8 \text{ kmol.}$$

C balance: amount of gas leaving second stage reactor is

$$(7 \cdot 07 \times 100/39 \cdot 8) = 17 \cdot 8 \text{ kmol}/100 \text{ kg feed}$$

N_2 balance:

$(N_2 \text{ in air added}) + (N_2 \text{ in gas ex-first stage}) = (N_2 \text{ in gas ex-second stage})$

$$y + 0 = (17 \cdot 8 \times 10 \cdot 3/100), \therefore \ y = 1 \cdot 83 \text{ kmol}/100 \text{ kg feed.}$$

Air is 79% v/v N_2, thus air added is

$$(1 \cdot 83 \times 100/79) = 2 \cdot 32 \text{ kmol}/100 \text{ kg feed}$$

Taking the mean molecular mass of air as $28 \cdot 9$ (Example 2-3),

$$\text{mass of air added} = (2 \cdot 32 \times 28 \cdot 9) = 67 \cdot 0 \text{ kg}/100 \text{ kg feed}$$
$$\text{or } 0 \cdot 67 \text{ kg/kg feed}$$

(c) In the gas leaving the CO converter, kmol C is

$$16.1 \text{ (in } CO_2) + 1.8 \text{ (in CO)} + 22.0 \text{ (in } CH_4) = 39.9 \text{ kmol}$$

$\therefore$ Amount of gas leaving CO converter is

$$(7 \cdot 07 \times 100/39.9) = 17 \cdot 7 \text{ kmol}/100 \text{ kg feed.}$$

The nitrogen in the gas leaving the converter = nitrogen in the gas leaving the CO_2 absorber, $\therefore$ amount of gas leaving the CO_2 absorber is

$$(0 \cdot 013 \times 17.7/0.165) = 11 \cdot 1 \text{ kmol}/100 \text{ kg feed,}$$
$$\text{or } (11 \cdot 1 \times 22 \cdot 4/100) = 2 \cdot 49 \text{ m}^3/\text{kg feed at STP.}$$

Example 2-7: A three component mixture containing 18% X, 31% Y and 51% Z is to be separated in a two stage distillation process; the bottom product from the first stage being used as feed to the second column. The maximum concentration of Y in the overhead product from the first column is to be 4%, and 75% of X in the feed will be recovered in this stream. The overhead product from the second column is to contain not more than 3% Z, and 75% of Y in the feed will be recovered in this stream. Finally, the bottom product from the second column should contain not more than 1% X and 80% of Z in the feed to the plant. What is the composition of each stream leaving the plant?

Solution
A simplified flow diagram is shown in Fig. 2-5

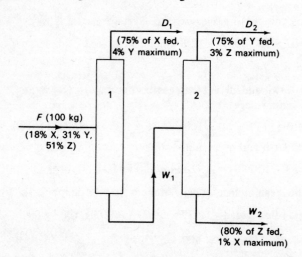

Fig. 2-5

Basis: 100 kg feed to the plant.

Overall balance $F = D_1 + D_2 + W_2$

Making the same balance for each component:
Component X

$$(100 \times 18/100) = (18 \times 75/100) + x_{D_2} \cdot D_2 + W_2/100$$

where x_{D_2} is the mass fraction of X in stream D_2

therefore

$$x_{D_2} \cdot D_2 + 0 \cdot 01 \, W_2 = 4 \cdot 5 \tag{2-1}$$

Component Y

$$(100 \times 31/100) = (4 \, D_1/100) + (31 \times 75/100) + y_{W_2} \cdot W_2$$

where y_{W_2} is the mass fraction of Y in stream W_2

therefore

$$0 \cdot 04 \, D_1 + y_{W_2} \cdot W_2 = 7 \cdot 75 \tag{2-2}$$

Component Z

$$(100 \times 51/100) = z_{D_1} \cdot D_1 + 3 \, D_2/100 + 51 \times 80/100$$

where z_{D_1} is the mass fraction of Z in stream D_1

therefore,

$$z_{D_1} \cdot D_1 + 0 \cdot 03 \, D_2 = 10 \cdot 2 \tag{2-3}$$

The sum of the mass fractions in any one stream, D_1, D_2 or W_2, must be unity and hence three further equations may be obtained as follows:

Stream D_1

$$(18 \times 75)/100 \, D_1) + 4 \, D_1/(100 \, D_1) + z_{D_1} \cdot D_1/D_1 = 1 \cdot 0$$

therefore

$$z_{D_1} = 0 \cdot 96 - 13 \cdot 5/D_1 \tag{2-4}$$

Stream D_2

$$x_{D_2} \cdot D_2/D_2 + (31 \times 75)/(100 \, D_2) + 3 \, D_2/(100 \, D_2) = 1 \cdot 0$$

therefore

$$x_{D_2} = 0 \cdot 97 - 23 \cdot 25/D_2 \tag{2-5}$$

Stream W_2

$$W_2/(100 \, W_2) + y_{W_2} \cdot W_2/W_2 + (51 \times 80)/(100 \, W_2) = 1 \cdot 0$$

therefore

$$y_{W_2} = 0 \cdot 99 - 40 \cdot 8/W_2 \tag{2-6}$$

Substituting for z_{D_1}, x_{D_2} and y_{W_2} from eqns. (2-4), (2-5) and (2-6) in eqns. (2-1), (2-2) and (2-3)

29

$$D_2 = 28 \cdot 3 - 0 \cdot 0103 \, W_2$$
$$W_2 = 49 \cdot 1 - 0 \cdot 0404 \, D_1$$
$$D_1 = 24 \cdot 7 - 0 \cdot 0312 \, D_2$$

Solving simultaneously

$$D_1 = 23 \cdot 83 \text{ kg}, \quad D_2 = 27 \cdot 80 \text{ kg and } W_2 = 48 \cdot 13 \text{ kg}$$

(a total of $99 \cdot 76$ kg, which compares with the feed of 100 kg indicating the accuracy of the calculations).

Substituting in eqns. (2-4), (2-5) and (2-6)

$$z_{D_1} = 0 \cdot 39$$
$$x_{D_2} = 0 \cdot 13$$

and

$$y_{W_2} = 0 \cdot 14$$

Therefore the composition (by mass) of each stream is

Stream D_1

X = 100(18 x 75)/(100 x 23·83)	= 56·7%	
Y = (4 x 23·83) 100/(100 x 23·83)	= 4·0%	
Z = (0·39 x 100)	= 39·0%	(Total 99·7%)

Stream D_2

X = (0·13 x 100)	= 13·0%	
Y = (31 x 75)100/(100 x 27·80)	= 83·6%	
Z = (3 x 27·80)100/(100 x 27·80)	= 3·0%	(Total 99·6%)

Stream W_2

X = (100 x 48·13)/(100 x 48·13)	= 1·0%	
Y = (0·14 x 100)	= 14·0%	
Z = (51 x 80)100/(100 x 48·13)	= 84·8%	(Total 99·8%)

2.3 Mass balance at unsteady-state

In section 2.1, an unsteady-state operation was defined as one in which there is a finite rate of accumulation (or depletion) within a plant or plant item. In general, there are two types of situation in which this occurs; time dependent operations and position dependent operations. The first of these includes problems relating to, for example, the filling of tanks, with or without an offtake, and batch distillation. In the latter operation, the compositions of both the material in the still and the vapour vary with time. The batch reactor provides a good example of a chemical system where composition is a function of time, and the solution of the problems involved here requires a knowledge of the reaction kinetics in the system. Examples of position dependent operations are continuous distillation and absorption within a column, where the composition of both streams varies with position in the column. Problems relating to this situation form the major part of chapters 7 and 8 and the discussion in the present section is therefore limited to time dependent operations.

In section 2.1, the basic relationship for mass balances was stated as

(mass in) = (mass out) + (accumulation)

or, in terms of rates,

(rate of input) = (rate of output) + (accumulation)

$$R_i = R_o + dM/dt \qquad (2\text{-}7)$$

where M kg is the mass of material within the plant or processing unit and R_i kg/s and R_o kg/s the flowrates of material into and out of the plant respectively – both these rates may vary with time. Rewriting eqn. (2-7) as a balance for one component in the system, i.e. in terms of concentration.

$$R_i C_i = R_o C + d(MC)/dt \qquad (2\text{-}8)$$

where C_i kg/kg is the concentration of the component in the feed stream – this may be a function of time, and C kg/kg is the concentration of the component in the plant at time t s. The last term in eqn. (2-8) may be expanded

$$d(MC)/dt = C dM/dt + M dC/dt \qquad (2\text{-}9)$$

From eqn. (2-7)

$$dM/dt = R_i - R_o$$

and therefore, in eqn. (2-8),

$$M dC/dt + R_i(C - C_i) = 0 \qquad (2\text{-}10)$$

The general solution of this equation depends on the forms of the functions M, R_i and C_i with time and may be expressed as

$$C = K \exp(- \int(R_i/M)dt) + \exp(-\int(R_i/M)dt) + (R_i C_i/M)$$
$$\times \exp(\int(R_i/M)dt) \, dt \qquad (2\text{-}11)$$

where K is the integration constant, usually evaluated from the conditions when $t = 0$. Calculations involving this equation are greatly simplified however, as, generally, one or more of the parameters is constant. Considering three important cases:

(a) If the inlet flow is equal to the outlet flow, then $dM/dt = 0$. If, in addition, $R_i = R_o$ = constant (say R), then it is usual to write

$R/M = 1/\theta$, where θ is the residence time (s).

(b) If, as well as constant feed and offtake rates, the inlet concentration, C_i is also a constant, then eqn. (2-10) becomes

$$M d(C - C_i)/dt + R(C - C_i) \quad = 0$$

or

$$d(C - C_i)/dt + (C - C_i)/\theta \quad = 0$$

This may be integrated to give

$$(C - C_i) = K \exp(-t/\theta) \qquad (2\text{-}12)$$

(c) Where the above conditions apply and, in addition, the concentration of the component in the inlet stream is zero, i.e. pure solvent, then $C_i = 0$, and

$$C = K \exp(-t/\theta) \qquad (2\text{-}13)$$

Example 2-8: A tank is filled with 450 kg of a salt solution containing $0 \cdot 05$ kg salt/kg solution which is perfectly mixed by vigorous agitation. Fresh water is added at a rate of $0 \cdot 19$ kg/s and the overflow is pumped to an identical second tank which is filled with solution, initially containing $0 \cdot 025$ kg salt/kg solution. After what time is the concentration of salt in the second tank a maximum?

Solution   For tank A, the feed is fresh water, so $C_i = 0$, and eqn. (2-13) may be used

$$C = K e^{-t/\theta}$$

when $t = 0$, $C = 0 \cdot 05$ kg/kg and hence $K = 0 \cdot 05$. The concentration in tank A as a function of t is therefore

$$C = 0 \cdot 05 \, e^{-t/\theta} \text{ kg/kg}$$

      For tank B, the inlet concentration, C_i is equal to that in the outlet stream from A:

$$C_i = 0.05 \, e^{-t/\theta}$$

Thus, in eqn. (2-10)

$$M dC/dt + R(C - 0.05 \, e^{-t/\theta}) = 0$$

or

$$dC/dt + C/\theta = 0.05 \, e^{-t/\theta}/\theta$$

Multiplying by $e^{t/\theta} \, dt$

$$e^{t/\theta} dC + C \, e^{t/\theta} dt/\theta = 0.05 \, dt/\theta$$

The left-hand side of this equation is the derivative of $C \, e^{t/\theta}$ and therefore

$$C \, e^{t/\theta} = \int 0.05 \, dt/\theta$$
$$= 0.05 \, t/\theta + K'$$

When $t = 0$, $C = 0.025$ kg/kg and hence $K' = 0.025$. The concentration of salt in tank B at time t is therefore

$$C = (0.05 \, t/\theta + 0.025) \, e^{-t/\theta} \text{ kg/kg}$$

Differentiating this equation

$$dC/dt = (0.05 \, e^{-t/\theta} - 0.05t \, e^{-t/\theta}/\theta)/\theta - 0.025 \, e^{-t/\theta}/\theta$$

The concentration of salt in tank B is a maximum when $dC/dt = 0$, i.e. when $t/\theta = 0.5$

In the present situation, M is a constant and equals 450 kg for both tanks. Similarly, $R_i = R_o = 0.19$ kg/s for both tanks; again a constant value.

$$\therefore \quad 1/\theta = R/M = (0.19/450)$$
$$= 0.000\,42 \text{ 1/s}$$
$$\therefore \quad 0.000\,42t = 0.5$$
$$\text{and } t = 1190 \text{ s}$$

2.4 Energy balances

As with mass, energy is conserved in any plant or operation, though in this case conversion between various forms of energy is an additional factor. Ignoring electrical energy for the moment, mechanical and thermal energy are of the greatest significance in chemical processes. Mechanical energy and its conservation in a system forms the basis of fluid mechanics which is dealt with in chapter 3 and is not therefore, considered here. An equally important consideration is that of energy balance relating to conservation of heat, where again, two situations are of relevance – steady-state and unsteady-state heat balances. The latter are discussed in chapters 4 and 6 and hence only steady-state heat balances are considered here.

2.5 Heat balance at steady-state

2.5.1 Physical systems

The basic statement for conservation of heat is

(enthalpy of products) – (enthalpy of feed streams) = (heat added to the process).

The enthalpy of the products, and similarly with the feeds, refers to the *sum* of all output enthalpies the difference between them may be negative when heat is withdrawn from the process. The heat added to (or withdrawn from) the process is the *sum* of the heat from all possible sources. These include heaters and coolers, chemical reactions, and enthalpy changes accompanying physical operations such as dilation, absorption, adsorption and so on.

In calculating enthalpies, it is vital that the enthalpies of all streams (and any reaction) are based on the same reference state and base temperature; both being important. For example, steam tables are based on 273 K and liquid water rather than solid ice. The choice of a sensible datum can do much to simplify calculations, especially if several streams enter the process at ambient temperature. If this temperature is selected as the datum, the enthalpies of the streams are zero and the amount of calculation is reduced considerably.

Example 2-9:   A flow of 1 kg/s air at 297 K is to be heated in a shell and tube unit using saturated steam at 136 kN/m^2. The flow of steam is 0·01 kg/s and the condensate, still at 136 kN/m^2, leaves the unit at 361 K via a steam trap. If the mean specific heat of air over the temperature range involved is 1·005 kJ/kg K, what is the exit air temperature? What would be the temperature of the exit gas if the steam was injected directly into the air stream? Neglect heat losses in both cases.

Solution

Shell and tube unit: From steam tables, enthalpy of saturated steam at 136 kN/m^2 is 2689 kJ/kg, enthalpy of water at 361 K is 368 kJ/kg.

Heat lost by steam is

$(2689 - 368) = 2321$ kJ/kg

$= (0\cdot01 \times 2321)$ kJ/s $= 23\cdot2$ kW

If the outlet air temperature is T K, then a balance may be made as follows. The datum will be taken as 297 K and hence, enthalpy of inlet air is 0. Enthalpy of outlet air is

$1\cdot005 \ (T - 297)$ kJ/kg

Therefore, heat gained by air is

$1\cdot005 \ (T - 297)$ kJ/kg $= 1 \times 1\cdot005 \ (T - 297)$ kJ/s
$= 1\cdot005T - 298\cdot5$ kW

Heat balance

(heat lost by steam) = (heat gained by air) + (losses)
$23\cdot2 = 1\cdot005T - 298\cdot5 + 0$
$\therefore \quad T = (321\cdot7/1\cdot005) = 320$ K.

Direct injection Taking a datum temperature of 273 K, heat in:

air $\quad = 1 \times 1\cdot005 \ (297 - 273) = 24\cdot1$ kW

steam $= (0\cdot01 \times 2689) = 26\cdot9$ kW

total heat into the system $= 51\cdot0$ kW

Heat out: assuming the air-steam mixture leaves at T K and 101·3 kN/m^2, the air-steam mixture will contain 1 kg air and 0·01 kg steam, that is

$(1/28\cdot9) = 0\cdot0346$ kmol air $\left.\begin{array}{c} \\ \\ \end{array}\right\}$ total $= 0\cdot035\ 16$
$(0\cdot01/18) = 0\cdot000\ 56$ kmol steam

Therefore, % volume of water vapour is

$(0\cdot000\ 56 \times 100/0\cdot035\ 16) = 1\cdot59$

Therefore, vapour pressure of water vapour is

$(1\cdot59 \times 101\cdot3/100) = 1\cdot61$ kN/m^2

From tables (chapter 10), the dew point is 287 K at which temperature the latent heat of water is 2466 kJ/kg. (The fact that the dew point is lower than the inlet air temperature rules out the latter as the datum in this case and 273 K is a wiser choice.)

Thus,

> heat in steam = (sensible heat in vapour) + (latent heat) + (sensible heat in water)

(assuming a mean specific heat of water vapour of $1·88$ kJ/kg K)

$$\text{heat in steam} = 0·01 \, (1·88 \, (T - 287) + 2466 + 4·187 \, (287 - 273))$$
$$= 0·0188T + 19·85 \text{ kW}$$

$$\text{heat in air} = 1 \times 1·005 \, (T - 273)$$
$$= 1·005T - 274·4 \text{ kW}$$

Therefore, total heat out of system (neglecting losses) is $1·0238T - 254·6$ kW

Heat balance:

$$51·0 = 1·0238T - 254·6$$

therefore

$$T = 298·5 \text{ K}$$

(In this second case, the rise in temperature of the air is $1·5$ K compared with 23 K where the heat is transferred indirectly, that is through a barrier. This is because the latent heat in the steam is not recovered, illustrating the advantages of the first case.)

Example 2-10: Flue gas from an oil-fired furnace passes to a waste-heat boiler in which steam is raised at 1170 kN/m^2. The following data were obtained during a test run:

> duration of trial $= 2·88$ ks
>
> composition of fuel $= 85\%$ C, 15% H
>
> consumption of fuel $= 18598$ kg
>
> CO_2 in flue gas entering the boiler $= 9:35\%$
>
> mean specific heat of wet flue gas $= 1·27$ kJ/m^3 K (at the flue gas temperature)
>
> temperatures: gas in WHB   1050 K   feed water   339 K
>
>   gas ex WHB   575 K   steam   459 K
>
> latent heat of steam at 459 K $= 1990$ kJ/kg
>
> steam used for oil atomization $= 0·3$ kg/kg

overall efficiency of WHB   = 87%

What is the rate of steam generation?

Solution

100 kmol dry flue gas contain 9·35 kmol C (as CO_2)
100 kg oil contain 85 kg C which is equivalent to (85/12) = 7·08 kmol C.

C balance: dry flue gas produced is

(7·08 x 100/9·35) = 75·7 kmol/100 kg fuel.

H_2 *balance:* 100 kg oil contain 15 kg H_2 which is equivalent to 15/2 = 7·5 kmol H_2. This produces 7·5 kmol water according to the equation

$$H_2 + \tfrac{1}{2}O_2 \rightarrow H_2O$$

In addition, 0·3 kg steam/kg oil (30 kg/100 kg oil) are used for atomization. 30 kg steam are equivalent to (30/18) = 1·73 kmol water vapour. Therefore the flue gas contains

(1·73 + 7·5) = 9·2 kmol water vapour.

Total of wet flue gas is

(9·2 + 75·7) = 84·9 kmol/100 kg oil
= (84·9 x 22·4/100) = 19·0 m³/kg fuel at 273 K

Heat balance (basis 1 kg oil burned).
Heat in:
Volume of wet flue gas at 1050 K is

(19·0 x 1050/273) = 73·1 m³

Taking a datum of 273 K, heat in wet flue gas is

73·1 x 1·27 (1050 − 273) = 72134 kJ/kg oil

Heat in feed water is

m x 4·187 (339 − 273) = 276·3 m kJ/kg oil

where m kg/kg oil is the rate of steam generation. Hence, total heat is (72134 + 276·3 m) kJ/kg oil.

Heat out: 87% of the heat into the WHB is transferred to the steam; the rest being lost by radiation or unaccounted for. Thus the losses are

13(72 134 + 276·3 m)/100 = 9377 + 35·92 m kJ/kg oil

Volume of wet flue gas at 575 K is

$19\cdot0 \times 575/273 = 40\cdot0$ m^3

Heat in wet flue gas is

$40\cdot0 \times 1\cdot27\,(575 - 273) = 15342$ kJ/kg oil

Heat in saturated steam at 459 K is

$1990\,m + m \times 4\cdot187\,(459 - 273) = 2768\cdot8\,m$ kJ/kg oil

Hence, total heat out is

$(24719 + 2804\cdot7\,m)$ kJ/kg oil.

Heat balance:

$$(72134 + 276\cdot3\,m) = (24719 + 2804\cdot7\,m)$$
$$\therefore \quad m = 18\cdot75 \text{ kg/kg oil}$$

Steam required for atomization is $0\cdot3$ kg/kg oil and therefore net output of steam is

$(18\cdot75 - 0\cdot3) = 18\cdot45$ kg/kg oil.

Oil used during trial is

$(18\ 598/2\cdot88) = 6458$ kg/ks or $6\cdot46$ kg/s

Rate of steam generation is

$(6\cdot46 \times 18\cdot45) = 119\cdot2$ kg/s

2.5.2 Chemical systems

Where a chemical reaction is involved, the heat passing into the system is not only the enthalpy of the inlet streams plus any heat supplied by a physical means, but also the heat given out (or taken in) by the reaction. This may be calculated from the standard heats of formation or combustion, which are always stated assuming that the formation or combustion takes place at 298 K. The heat of reaction thus assumes that the reaction takes place also at 298 K and that the reactants and products enter and leave the system at this temperature. This is rarely the case and it is usually necessary to calculate the true enthalpy of a stream after assuming reaction at 298 K. In this way, there is no choice of datum temperature and this, together with the phases of the reactants and products, is fixed by the heat of formation data.

Example 2-11: In the manufacture of sulphuric acid, SO_2 is oxidized to SO_3 in two catalytic converters with intermediate cooling. The feed to the first converter, in which 80% of the SO_2 is converted to SO_3, enters at 698 K and has the following analysis by volume: 9·0% SO_2, 10·1% O_2 and 80·9% N_2. Based on the following data, at what temperature does the gas stream enter the intermediate cooler? (Heats of formation at 298 K : $SO_2(g) = -297\,000$ kJ/kmol, $SO_3(g) = -394\,000$ kJ/kmol $O_2(g) = 0$. Specific heat as a function of T K:

$$N_2 = 27\cdot1 + 5\cdot81 \times 10^{-3}T - 0\cdot29 \times 10^{-6}T^2 \qquad kJ/kmol\ K$$

$$O_2 = 25\cdot6 + 13\cdot28 \times 10^{-3}T - 4\cdot18 \times 10^{-6}T^2 \qquad kJ/kmol\ K$$

$$SO_2 = 29\cdot1 + 41\cdot8 \times 10^{-3}T - 15\cdot75 \times 10^{-6}T^2 \qquad kJ/kmol\ K$$

$$SO_3 = 31\cdot2 + 80\cdot2 \times 10^{-3}T - 27\cdot7 \times 10^{-6}T^2 \qquad kJ/kmol\ K)$$

Solution
Mass balance around the first converter (basis 100 kmol feed to the plant): In 100 kmol feed, 9·0 kmol SO_2, 10·1 kmol O_2 and 80·9 kmol N_2, 80% of the SO_2 is converted according to the reaction

$$SO_3 + \tfrac{1}{2}O_2 \rightarrow SO_3$$

Thus, $(80 \times 9\cdot0/100) = 7\cdot2$ kmol react, consuming $(7\cdot2/2) = 3\cdot6$ kmol O_2 and forming 7·2 kmol SO_3. The gas leaving the first stage converter therefore contains

$$SO_2 = (9\cdot0 - 7\cdot2) = 1\cdot8 \text{ kmol}$$
$$O_2 = (10\cdot1 - 3\cdot6) = 6\cdot5 \text{ kmol}$$
$$SO_3 = 7\cdot2 \text{ kmol}$$
$$N_2 = 80\cdot9 \text{ kmol (unchanged)}$$

Heat of reaction at 298 K:
formation of SO_2: $\qquad S + O_2 = SO_2 + 297\,000$ kJ/kmol SO_2 formed

formation of SO_3: $\qquad S + 3/2\,O_2 = SO_3 + 394\,000$ kJ/kmol SO_3 formed

The reaction under consideration is

$$SO_2 + \tfrac{1}{2}O_2 = SO_3 + x \qquad\qquad kJ/kmol\ SO_2 \text{ reacting}$$

Thus, substituting from the two formation equations

$$S + O_2 - 297\,000 + \tfrac{1}{2}O_2 = S + 3/2\,O_2 - 394\,000 + x$$

$$\therefore \quad x = 97\,000 \text{ kJ/kmol } SO_2 \text{ reacting}$$

For 100 kmol feed, 7·2 kmol SO_2 reacts and hence heat evolved due to the reaction is

$$(7·2 \times 97\,000) = 698\,000 \text{ kJ}$$

Heat balance around the first converter (datum 298 K, basis 100 kmol feed): Consider the heat in

The mean specific heat of the feed gas, c_{p_m} between temperatures T_1 T_2 K is given by

$$c_{p_\mathrm{m}} = \int_{T_1}^{T_2} c_p \, dT / (T_2 - T_1) \tag{2-14}$$

The enthalpy of each component, H_i is given by

$$H_i = n_i c_{p_{\mathrm{m}_i}} (T_2 - T_1)$$

where n_i is the kmol of component i

Therefore

$$H_i = n_i \left(\int_{T_1}^{T_2} c_p \, dT \right)$$

$$= n_i \int_{T_1}^{T_2} (a_i + b_i T + c_i T^2) \, dT$$

a, b and c being constants as quoted in the question. Therefore

$$H_i = n_i (a_i T_2 + \tfrac{1}{2} b_i T_2^{\,2} + \tfrac{1}{3} c_i T_2^3 - a_i T_1 - \tfrac{1}{2} b_i T_2^2 - \tfrac{1}{3} c_i T_2^3)$$

Total enthalpy, (2-15)

$$H = \Sigma H_i = \Sigma n_i a_i T_2 + \tfrac{1}{2} \Sigma n_i b_i T_2^2 + \tfrac{1}{3} \Sigma n_i c_i T_2^3$$
$$- \Sigma n_i a_i T_1 - \tfrac{1}{2} \Sigma n_i b_i T_1^2 - \tfrac{1}{3} \Sigma n_i c_i T_1^3 \tag{2-16}$$

Tabulating data for the feed stream

i	n_i	$n_i a_i$	$n_i b_i$	$n_i c_i$
SO_2	9·0	261·9	$376·2 \times 10^{-3}$	$-141·8 \times 10^{-6}$
O_2	10·1	258·5	$134·0 \times 10^{-3}$	$-42·3 \times 10^{-6}$
N_2	80·9	2190·0	$470·0 \times 10^{-3}$	$-23·5 \times 10^{-6}$
		$\Sigma n_i a_i = 2710·4$	$\Sigma n_i b_i = 980·2 \times 10^{-3}$	$\Sigma n_i c_i = -207·6 \times 10^{-6}$
			$\tfrac{1}{2}\Sigma n_i b_i = 490·1 \times 10^{-3}$	$\tfrac{1}{3}\Sigma n_i c_i = -69·2 \times 10^{-6}$

Heat in feed gas is

$$(2710 \cdot 4 \times 698) + (490 \cdot 1 \times 10^{-3} \times 698^2) - (69 \cdot 2 \times 10^{-6} \times 698^3)$$
$$- (2710 \cdot 4 \times 298) - (490 \cdot 1 \times 10^{-3} \times 298^2) + (69 \cdot 2 \times 10^{-6} \times 298^3)$$

$$= 1\ 258\ 200\ \text{kJ}/100\ \text{kmol feed.}$$

Thus,

total heat into first converter = feed + reaction

$$= 1\ 258\ 200 + 698\ 000$$
$$= 1\ 956\ 000\ \text{kJ}/100\ \text{kmol feed.}$$

Consider the heat out: Ignoring losses, the only heat out is that in the product stream leaving the first reactor, whose temperature is unknown. Taking this as T K, for the product stream

i	n_i	$n_i a_i$	$n_i b_i$	$n_i c_i$
SO_2	1·8	52·4	$75 \cdot 4 \times 10^{-3}$	$-28 \cdot 4 \times 10^{-6}$
O_2	6·5	167·2	$86 \cdot 3 \times 10^{-3}$	$-27 \cdot 2 \times 10^{-6}$
SO_3	7·2	209·5	$301 \cdot 0 \times 10^{-3}$	$-113 \cdot 2 \times 10^{-6}$
N_2	80·9	2190·0	$470 \cdot 0 \times 10^{-3}$	$-23 \cdot 5 \times 10^{-6}$
		$\Sigma n_i a_i = 2619 \cdot 1$	$\Sigma n_i b_i = 932 \cdot 7 \times 10^{-3}$	$\Sigma n_i c_i = -192 \cdot 3 \times 10^{-6}$
			$\tfrac{1}{2}\Sigma n_i b_i = 466 \cdot 4 \times 10^{-3}$	$\tfrac{1}{3}\Sigma n_i c_i = -64 \cdot 1 \times 10^{-6}$

Heat in exit gas is

$$(2618 \cdot 1T) + (466 \cdot 4 \times 10^{-3} T^2) - (64 \cdot 1 \times 10^{-6} T^2) - (2619 \cdot 1 \times 298)$$
$$- (466 \cdot 4 \times 10^{-3} \times 298^2) + (64 \cdot 1 \times 10^{-6} \times 298^3)$$

$$= (2619 \cdot 1T + 466 \cdot 4 \times 10^{-3} T^2 - 64 \cdot 1 \times 10^{-6} T^3)$$
$$- 819\ 300\ \text{kJ}/100\ \text{kmol feed.}$$

Heat balance

$$1\ 956\ 200 = 2619 \cdot 1T + 466 \cdot 4 \times 10^{-3} T^2 - 64 \cdot 1 \times 10^{-6} T^3 - 819\ 300$$

or

$$2619 \cdot 1T + 0 \cdot 466T^2 - 64 \cdot 1 \times 10^{-6} T^3 = 2\ 775\ 500$$

This equation is solved perhaps most easily by graphical means as shown in Fig. 2-6, where values of the function have been calculated for

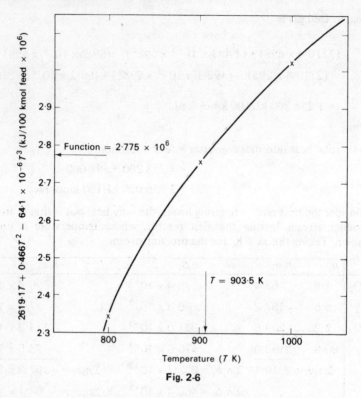

Fig. 2-6

$T = 800$, 900 and 1000 K. From the figure, it will be seen that an enthalpy balance is obtained when $T = 903 \cdot 5$, say 904 K.

(This problem illustrates the situation in which specific heat data are available as a function of temperature. The use of mean values greatly simplifies the working, however, and will suffice for many engineering calculations, especially where a limited temperature range is involved.)

2.6 Simultaneous mass and heat balances

Although several of the problems in this chapter have involved both mass and heat balances, the general procedure has been to complete the mass balance before going on to the calculation of enthalpies. There are many cases, however, especially in distillation calculations (chapter 7), where it is not possible to complete the mass balance, and simultaneous solution is necessary.

Example 2-12: Quicklime is produced by heating limestone with pure coke in a vertical kiln, through which air is blown. In a test run, during which the limestone, coke and air entered the kiln at 298 K, the quicklime, which contained no carbon or limestone, was removed from the bottom of the kiln at 783 K. The exit gas left the top of the kiln at 588 K and this contained carbon dioxide and nitrogen only. Ignoring heat losses from the unit, what was the ratio of limestone to coke fed by mass? (Standard heats of formation at 298 K

$$CO_2 (g) = -398\ 000\ kJ/kmol$$
$$CaO(s) = -635\ 000\ kJ/kmol$$
$$CaCO_3 (s) = -1\ 211\ 000\ kJ/kmol$$

Mean specific heats

$$CaO(s) = 51 \cdot 9\ kJ/kmol\ K$$
$$CO_2 = 49 \cdot 0\ kJ/kmol\ K$$
$$N_2 = 29 \cdot 3\ kJ/kmol\ K)$$

Solution *Mass balance* (basis: 100 kmol limestone fed).

A simplified flow diagram is shown in Fig. 2-7. The two reactions involved are

$$CaCO_3 \rightarrow CaO + CO_2$$
$$C + O_2 \rightarrow CO_2$$

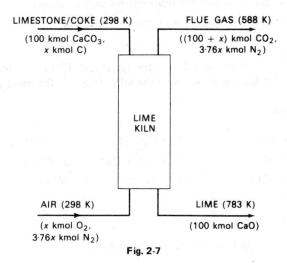

LIMESTONE/COKE (298 K)

(100 kmol $CaCO_3$, x kmol C)

FLUE GAS (588 K)

((100 + x) kmol CO_2, 3·76x kmol N_2)

LIME KILN

AIR (298 K)

(x kmol O_2, 3·76x kmol N_2)

LIME (783 K)

(100 kmol CaO)

Fig. 2-7

43

CaO balance:
100 kmol $CaCO_3$ contain 100 kmol CaO which are converted to CaO and leave as lime. This stream is pure lime, and hence amount of lime produced is 100 kmol

C balance:
100 kmol $CaCO_3$ contain 100 kmol C which form 100 kmol CO_2 x kmol coke contain x kmol C which form x kmol CO_2. Therefore, total CO_2 in flue gas is $(100 + x)$ kmol.

O_2 balance:

$$(O_2 \text{ in air}) + (O_2 \text{ in limestone}) = (O_2 \text{ in lime}) + (O_2 \text{ in flue gas})$$
$$y \quad + (3 \times 100/2) \quad = (100/2) \quad + (100 + x)$$
$$y = x \text{ kmol.}$$

N_2 balance:
Air supplied is $(100x/21) = 4\cdot76\,x$ kmol and this contains $(79 \times 476x/100) = 3\cdot76x$ kmol N_2.

This leaves with flue gas and hence the flue gas contains $3\cdot76$ kmol N_2. These data are summarized in Fig. 2-7. It is not possible to solve for x at this stage and the heat balance must now be made.

Heat balance (datum: 298 K, basis: 100 kmol $CaCO_3$ fed).

$$\text{(heat in)} + \text{(heat of reaction)} = \text{(heat out)} + \text{(losses)}$$

Considering each of these terms in turn

Heat in:
Both feeds enter the kiln at the datum 298 K, i.e. the limestone/coke and the air. The enthalpy of the inlet streams is therefore zero.

Heat of reaction:

formation of $CaCO_3$: $\quad Ca + C + \frac{3}{2}O_2 = CaCO_3 + 1\,211\,000$ kJ
formation of CaO $\quad$: $\quad Ca + \frac{1}{2}O_2 = CaO + 635\,000$ kJ
formation of CO_2 $\quad$: $\quad C + O_2 = CO_2 + 398\,000$ kJ

Therefore for the reaction

$$CaCO_3 = CaO + CO_2 + A \text{ kJ}$$

we can substitute from the formation reactions

$$Ca + C + \tfrac{3}{2}O_2 - 1\ 211\ 000 = Ca + \tfrac{1}{2}O_2 - 635\ 000 + C + O_2$$
$$- 398\ 000 + A \text{ kJ}$$

Therefore

$$A = -178\ 000 \text{ kJ/kmol } CaCO_3 \text{ converted.}$$

Thus for 100 kmol $CaCO_3$, 17 800 000 kJ must be supplied. This is attained by combustion of the coke

$$C + O_2 = CO_2 + 398\ 000 \text{ kJ/kmol}$$

Therefore, heat evolved by x kmol C is 389 000x kJ/100 kmol limestone, and net heat of reaction is

$$398\ 000x - 17\ 800\ 000 \text{ kJ/100 kmol limestone}$$

Heat out:

There are two exit streams from the plant, the lime and the flue gas. Considering these in turn.

Lime: enthalpy of exit stream is

$$100 \times 51 \cdot 9 (783 - 298) = 2\ 517\ 150 \text{ kJ/100 kmol limestone.}$$

Flue gas: enthalpy of N_2 is

$$3 \cdot 76x \times 29 \cdot 3 (588 - 298) = 31\ 949x \text{ kJ/100 kmol limestone.}$$

Also, enthalpy of CO_2 is

$$(100 + x) \times 49 \cdot 0 (588 - 298) = 14\ 210 (100 + x) \text{ kJ/100 kmol limestone.}$$

The losses may be neglected and hence the total heat out is

$$2\ 517\ 150 + 31\ 949x + 14\ 210 (100 + x) = 3\ 938\ 150 + 46\ 159x \text{ kJ}$$

Heat balance:

$$0 + 398\ 000x - 17\ 800\ 000 = 3\ 938\ 150 + 46\ 159x$$

From which, x is 61·8 kmol/100 kmol limestone. Mass of coke fed is

$$(61 \cdot 8 \times 12) = 741 \cdot 6 \text{ kg/100 kmol limestone}$$

Now, 100 kmol limestone is equivalent to

$$(100 \times 100) = 10\ 000 \text{ kg limestone}$$

Therefore, ratio of limestone/coke feed is

$$(10\ 000/741 \cdot 6) = 13 \cdot 5 \text{ kg/kg}$$

List of symbols

Most of the symbols are defined in the relevant problem, however, the more general nomenclature is listed as follows:

a, b, c	constants in the expression for sensible heat content as a function of temperature	$(-)$
c_p	sensible heat content	(kJ/kmol K)
C	concentration of a component at time t	(kg/kg)
C_i	concentration of a component in the inlet stream	(kg/kg)
D	mass flowrate of overheat product	(kg/s)
F	mass flowrate of feed stream(s)	(kg/s)
H	enthalpy	(kJ)
K, K'	constants of integration	$(-)$
M	mass of material within a plant or processing unit	(kg)
m	rate of steam generation/unit weight of fuel burned	(kg/kg)
n_i	amount of component i	(kmol)
R_i, R_o	mass flowrate of input and output streams respectively	(kg/s)
T	temperature	(K)
t	time	(s)
W	mass flowrate of bottom product	(kg/s)
θ	residence time	(s)

subscripts

i	referring to the ith component
i	referring to inlet stream
m	referring to a mean value

Further reading

-LEWIS, W. K. & RADASCH, A. H. *Industrial Stoichiometry.* McGraw-Hill New York (1959).

SCHMIDT, A. X. & LIST, H. L. *Material & Energy Balances.* Prentice-Hall New York (1962).

TYNER, M. *Process Engineering Calculations.* Ronald Press (1960).

WHITWELL, J. C. & TONER, R. K. *Conservation of Mass & Energy.* Blaisdell Publ. Co. U.S.A. (1969).

WILLIAMS, E. T. & JOHNSON, C. E. *Stoichiometry for Chemical Engineers.* McGraw-Hill New York (1958).

DAVIES, C. *Calculations in Furnace Technology.* Pergamon Press Oxford (1970).
PECK, W. J. & RICHMOND, A. J. *Applied Thermodynamics Problems for Engineers.* Edward Arnold London (1957).
HARKER, J. H. & ALLEN, D. A. *Fuel Science.* Oliver & Boyd Edinburgh (1972).
BACKHURST, J. R. & HARKER, J. H. *Process Plant Design.* Heinemann Educational Books London (1973)

Problem 2-1: A benzene and toluene mixture containing 40% benzene by weight is to be separated in a continuous distillation column. If the concentration of benzene by weight is to be 55% and 5% in the top and bottom products respectively, calculate:

(a) the amount of top product as a percentage of the total feed, and
(b) the percentage of benzene fed to the column which is recovered in the top product.

$$(70\%; 96\cdot25\%)$$

Problem 2-2: A gas of the following volumetric composition: CO_2 $3\cdot0\%$, CO $9\cdot0\%$, C_3H_6 $3\cdot0\%$, CH_4 $24\cdot0\%$, H_2 $53\cdot6\%$, O_2 $0\cdot4\%$, N_2 $6\cdot0\%$ is burned to produce a flue gas containing CO_2 $7\cdot2\%$, O_2 $7\cdot4\%$, N_2 $85\cdot4\%$ on a dry basis.
Calculate the theoretical and actual air supplied.

$$(4\cdot41 \text{ m}^3/\text{m}^3; 50\cdot6\% \text{ excess})$$

Problem 2-3: An oil containing carbon and hydrogen only produces a dry flue gas of the following volumetric composition: CO_2 $10\cdot7\%$, O_2 $7\cdot4\%$, N_2 $81\cdot9\%$. Calculate the composition of the fuel and the percentage excess air used in the combustion.

$$(89\cdot3\%C, 10\cdot7\%H \text{ by weight}; 51.6\% \text{ excess})$$

Problem 2-4: A flow, $0\cdot126$ kg/s, of sand containing 20% water is to be dried to produce a product containing 5% water. The partial pressure of water vapour in the ambient air is $1\cdot33$ kN/m^2 and the partial pressure of water vapour in the air leaving the dryer must not exceed $26\cdot7$ kN/m^2. In order to achieve this, a proportion of the exit air is recycled and mixed with the inlet air so that the partial pressure of water vapour in the air fed to the dryer is $6\cdot7$ kN/m^2. Assuming atmospheric pressure to be $101\cdot3$ kN/m^2, calculate:

(a) the mass of fresh air fed to the dryer per second, and
(b) the air recycle rate.

$$(0\cdot091 \text{ kg/s}; 0\cdot022 \text{ kg/s})$$

Problem 2-5:   A stream, 1 kg/s, containing 50% xylenes, 30% toluene and 20% benzene is fed to a continuous two-column distillation unit. The overheads from the first column contain $4\cdot5\%$ xylenes, $9\cdot1\%$ toluene and $86\cdot4\%$ benzene and the bottom product is used as feed to the second column. The products from the second column are:

overheads — $6\cdot9\%$ xylenes, $90\cdot1\%$ toulene, $3\cdot0\%$ benzene
bottoms   — $95\cdot1\%$ xylenes, 4·1 toluene, 0·4% benzene.

Calculate the flow of:

(a) the overhead product from the first column,
(b) the overhead product from the second column,
(c) the bottom product from the first column, and
(d) the bottom product from the second column.

$$(0\cdot22 \text{ kg/s}; 0\cdot29 \text{ kg/s}; 0\cdot49 \text{ kg/s}; 0\cdot78 \text{ kg/s})$$

Problem 2-6:   A well-stirred tank contains a solution of salt in water. Fresh water is added at such a rate that the residence time is 600 s. After what time is the concentration of salt in the tank reduced to 1% of its original value? What percentage of the original salt is removed when an amount of fresh water equal to the volume of the tank has been added?

$$(2\cdot77 \text{ ks}; 63\cdot1\%)$$

Problem 2-7:   Flue gas of the following composition by volume (dry basis): CO_2 $11\cdot6\%$, O_2 $6\cdot8\%$, CO $0\cdot5\%$ leaves a boiler at 538 K. If the inlet air temperature is 287 K and the fuel contains $79\cdot5\%$ carbon, estimate the heat carried away by the flue gas per kg fuel burned. (Mean specific heat of flue gas = $1\cdot005$ kJ/kg K).

$$(4230 \text{ kJ/kg fuel})$$

Problem 2-8:   In a producer gas plant, coke containing 84% C, $9\cdot5\%$ ash produces a gas of calorific value 5750 kJ/m^3 and the following volumetric composition: CO $23\cdot5\%$, CO_2 $4\cdot8\%$, H_2 $16\cdot2\%$, CH_4 $1\cdot8\%$, N_2 $53\cdot7\%$.

If the ash leaving the plant contains $12\cdot8\%$ C, estimate, for 1 kg coke fired:

    (a) the weight of carbon gasified
    (b) the volume of gas produced
    (c) the heat available in the gas.

Hint: (amount of C forming CO)/(total C burned) =
% CO/(%CO + %CO$_2$)
      (0·826 kg/kg coke; 5·12 m^3/kg coke at STP; 29 500 kJ/kg coke)

3. Basic Fluid Mechanics

3.1 Introduction

A detailed understanding of fluid mechanics is an essential basis for the subsequent study of heat and mass transfer processes. Whilst the discussion of inviscid fluids necessarily forms an important section of any course concerned with the behaviour of fluids, the problems in this chapter are restricted to those dealing with effects which result from the transfer of momentum, either at the molecular level (through the action of viscosity) or by turbulent phenomena.

3.2 Rheological properties of fluids

3.2.1 Newtonian and non-Newtonian fluids

Many liquids of interest to the process and chemical engineer do not exhibit Newtonian behaviour, that is, strict proportionality between shear stress τ and shear rate (rate of strain) $\dot{\gamma}$. The shear stress/shear rate characteristics (rheograms) for some common non-Newtonian liquids are shown in Fig. 3-1. Since the characteristic for the Newtonian liquid is (by definition) linear, we have the familiar relationship

$$\tau = \mu\dot{\gamma} = \mu\frac{dv}{dy} \quad \text{(Newton's Law of Viscosity)} \tag{3-1}$$

where the constant of proportionality, μ, is the viscosity of the liquid.

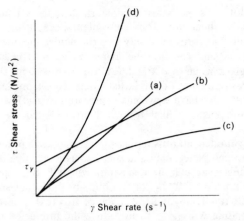

(a) Newtonian fluid: Viscosity $(\tau/\dot{\gamma})$ = constant.
(b) Bingham plastic: for $\tau > \tau_y$ liquid behaves in Newtonian manner.
(c) Pseudo plastic liquid: viscosity decreases with increasing shear rate
(d) Dilatant liquid: viscosity increases with increasing shear rate.

Fig. 3-1

Bingham plastics are liquids which exhibit a definite yield strength; below this stress no flow occurs. At stresses above the yield value the behaviour is Newtonian. The constitutive equation for the Bingham plastic is

$$\tau - \tau_y = \mu\dot{\gamma} = \mu\frac{\mathrm{d}v}{\mathrm{d}y} \tag{3-2}$$

Two other types of behaviour are shown in Fig. 3-1 namely shear thinning and shear thickening. Liquids which exhibit shear thinning, that is a decrease in viscosity with increasing shear rate, are said to be pseudo-plastic. Liquids which show an increase in viscosity with increasing shear rate are called dilatant liquids.

There are several useful constitutive equations for pesudoplastic and dilatant liquids, although they all have limited ranges of applicability. The most common equation is the empirical Ostwald-de Waele expression

$$\tau = \mu\dot{\gamma}^n = K\left(\frac{\mathrm{d}v}{\mathrm{d}y}\right)^n \tag{3-3}$$

more generally known as the power law equation. This form of equation is suggested by the observation that for many liquids a logarithmic plot of shear stress against shear rate is often linear over large ranges of shear rate. It is convenient because of its simple form, and the fact that it involves only two rheological parameters, the consistency index K and the flow behaviour index n. The larger the value of K at a given value of n and dv/dy, the more viscous the liquid. The value of n is a measure of the liquid's departure from Newtonian behaviour. For $n = 1$, we have Newtonian behaviour ($K = \mu$), with $n < 1$ behaviour is pseudoplastic, whilst for $n > 1$ eqn. 3-3 can be used for dilatant liquids.

The power law model can thus represent a whole range of behaviour; from dilatant, through Newtonian and pseudoplastic, to plug flow. The discussion of fluids that exhibit behaviour which is time dependent (rheopectic or thixotropic liquids), or combines both viscous and elastic effects (viscoelasticity) is beyond the scope of this text. However, the list of further reading will serve to introduce the interested reader to this subject.

Examples 3-1, 3-2, 3-3 and 3-4 demonstrate the measurement and subsequent use of rheological data, shear stress/shear rate characteristics, for flowrate/pressure drop determination.

In the simple one dimensional flows considered in this chapter $\dot{\gamma} = dv/dy$, this is not true for more complex flows.

Example 3-1: The rheological properties of many process liquids are conveniently measured by means of a viscometer based on the cone and plate principle. Show how measurements of shear rate and shear stress are achieved with this type of instrument.

A typical cone and plate viscometer has a cone with an angle of $0.5°$ and a diameter of 30 mm. When the cone rotates at 90 rpm (1.5 Hz), the torque applied at the cone shaft is 0.34×10^{-2} N m. If the liquid under test is known to be Newtonian, calculate its viscosity.

Solution: The essential elements of the cone and plate viscometer are shown in Fig. 3-2, although in this diagram the cone angle ϕ has been exaggerated for the sake of clarity.

Let the rotary speed of the cone be N Hz. then the angular velocity of the cone

$$\omega = 2\pi N \text{ radians/s}$$

Tangential velocity of the cone surface at radius $r = r\omega = 2\pi r N$

Depth of the liquid sample at radius r is

$$\delta = r \tan \phi \quad \text{(where } \phi \text{ is the cone angle)}$$

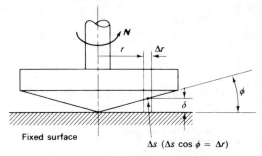

Fig. 3-2

Since δ is very small in a practical instrument, the velocity gradient experienced by the liquid at radius r

$$\doteqdot \frac{2\pi rN}{\delta}$$

Therefore, velocity gradient or shear rate $\dot{\gamma}$ is

$$\frac{2\pi rN}{r \tan \phi} = 2\pi N/\tan \phi$$

which is conveniently independent of radius.

Therefore a test liquid between the cone and the plate experiences an essentially uniform shear rate at any particular cone speed. Since shear rate is constant the shear stress over the surface of the cone will also be constant.

Consider a radial element of the cone at radius r. Shear force on the annular element at r is $2\pi r\Delta s\tau$ where τ is the shear stress. Since

$$\Delta r = \Delta s \cos \phi \quad \text{and} \quad \cos \phi \to 1 \quad \text{as} \quad \phi \to 0 \quad \text{then} \quad \Delta r \doteqdot \Delta s$$

Therefore, shear force on element at radius r is $2\pi r\Delta r\tau$. Torque due to shear force on the annular element is

$$2\pi r\Delta r\tau r = 2\pi r^2 \delta r\tau.$$

If the measured torque is T, then

$$T = \int_0^R 2\pi \tau r^2 \, dr = 2/3 \, \pi\tau R^3$$

Hence, shear stress τ is

$$\frac{3T}{2\pi R^3}$$

and shear rate $\dot{\gamma}$ is

$$\dot{\gamma} = \frac{2\pi N}{\phi}$$

since $\tan \phi \doteqdot \phi$ for small ϕ.

Therefore, direct measurement of T and N permits the calculation of τ and $\dot{\gamma}$. By definition

$$\text{viscosity } \mu = \frac{\text{shear stress}}{\text{shear rate}}.$$

A Newtonian liquid is one which exhibits a linear variation of shear stress with shear rate and therefore the ratio of shear stress/shear rate is a constant.

For a cone angle of $0 \cdot 5°$ ($= 0 \cdot 0087$ radians), shear rate is

$$\frac{2\pi(90/60)}{(0 \cdot 5/360)2\pi} = \frac{1 \cdot 5 \times 360}{0 \cdot 5} = 1080 \text{ 1/s}$$

Shear stress is

$$\frac{3 \times 0 \cdot 34 \times 10^{-2}}{2\pi(1 \cdot 5/100)^3} = 483 \text{ N/m}^2$$

The viscosity of the liquid is

$$\frac{\tau}{\dot{\gamma}} = \frac{483}{1080} = 0 \cdot 447 \text{ Ns/m}^2$$

Example 3-2: An incompressible Newtonian liquid flows, under laminar conditions, through a very long horizontal duct. If the width of the duct N is large in comparison with its depth $2h$, derive an expression which gives the velocity V of liquid as a function of the distance from the mid-plane of the duct.

A liquid with a viscosity of 3 Ns/m² flows through a duct having a depth of 10 mm. If the mean flow velocity through the duct is 50 mm/s calculate the shear stress to which the duct walls will be subjected. Use this parameter to calculate the axial pressure drop/m of the duct length.

Solution A section of the duct is shown in Fig. 3-3. If steady-state conditions prevail, and the flow is not influenced by duct entry conditions then an element will be subjected to forces shown, and these

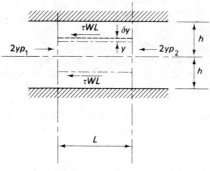

Fig. 3-3

forces will be in equilibrium, therefore

$$(P_1 - P_2)2yW = \tau\,2WL$$

where τ is the shear stress, with $\tau = f(y)$. Since the duct width W, assumed large in comparison with $2h$ shear forces on the sides of the element, have been neglected, then

$$\tau = (P_1 - P_2)\frac{y}{L} = \mu\,\frac{\mathrm{d}V}{\mathrm{d}y^*}$$

where y^* is the distance from the duct wall. Therefore,

$$y^* = h - y \qquad \frac{\mathrm{d}V}{\mathrm{d}y^*} = -\frac{\mathrm{d}V}{\mathrm{d}y}$$

Hence

$$(P_2 - P_1)y/\mu L = -\frac{\mathrm{d}V}{\mathrm{d}y}$$

Integration gives

$$V = -\frac{(P_1 - P_2)y^2}{2\mu L} + C$$

when $y = h$, $V = 0$, therefore

$$C = \frac{(P_1 - P_2)h^2}{2\mu L}$$

and the required velocity distribution becomes

$$V = \frac{(P_1 - P_2)h^2}{2\mu L} \left[1 - \left(\frac{y}{h} \right)^2 \right]$$

Since we are normally concerned with average velocities in practical design calculations, the velocity distribution determined above is not particularly useful as it stands.

The volumetric flowrate/unit width of duct, Q is

$$2 \int_0^h V \mathrm{d}y$$

where $V\mathrm{d}y$ is the volumetric flowrate through an elementary strip δy at a distance y from the duct centre line. Thus

$$Q = 2 \int_0^h \frac{(P_1 - P_2)h^2}{2\mu L} \left[1 - \left(\frac{y}{h} \right)^2 \right] \mathrm{d}y = \frac{(P_1 - P_2)2h^3}{3\mu L}$$

Average velocity is

$$\frac{Q}{2h} = \frac{(P_1 - P_2)h^2}{3\mu L} = V_A$$

The velocity distribution may be written

$$V = 3/2 \ V_A \left[1 - \left(\frac{y}{h} \right)^2 \right]$$

(when $y = 0$, $V = V_{max} = 1 \cdot 5 \ V_A$)

Shear rate at the wall is

$$\left(\frac{\mathrm{d}V}{\mathrm{d}y} \right)_{y=h} = - \left(\frac{3yV_A}{h^2} \right)_{y=h} = - \frac{3V_A}{h}$$

Shear stress at the wall

$$\tau_w = - \frac{3V_A \mu}{h}$$

The negative sign denotes that the direction of this stress is opposite to the direction of flow.

If $V_A = 50$ mm/s, $\mu = 3$ Ns/m^2, $h = 5$ mm, then

$$\tau_w = \frac{3 \times 5 \times 3}{0 \cdot 5} = 90 \ \text{N/m}^2$$

A force balance over the length of the duct gives

$$2(\tau_w WL) = 2hW(P_1 - P_2)$$

For $L = 1$ m

$$(P_1 - P_2) = \frac{\tau_w}{h} = \frac{90}{0\cdot5 \times 10^{-2}} = 18 \text{ kN/m}^2$$

Example 3-3: Rheological tests on a liquid provide data which indicate that its shear stress/shear rate characteristic can be represented accurately by the equation

$$\tau = -K \left| \frac{\mathrm{d}v}{\mathrm{d}y} \right|^{n-1} \frac{\mathrm{d}v}{\mathrm{d}y} \qquad \text{where } K \text{ and } n \text{ are constant}$$

in tube flow $\mathrm{d}v/\mathrm{d}r$ is everywhere negative and under these circumstances the above equation may be written as

$$\tau = K \left(-\frac{\mathrm{d}v}{\mathrm{d}r} \right)^n$$

Derive an expression for the velocity distribution which such a liquid would exhibit when flowing through a long tube of radius R. Also produce an equation relating volumetric flowrate and pressure drop along the tube. Comment on this result for $n = 1$.

Solution: A section of the tube is shown in Fig. 3-4. As before, steady state conditions are assumed and thus a force balance on a cylindrical element of liquid, radius r, gives

$$(P_1 - P_2)\pi r^2 = 2\pi rL\tau$$

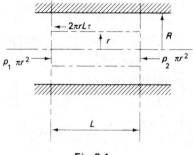

Fig. 3-4

Therefore

$$\left(-\frac{dv}{dr}\right)^n = \left(\frac{(P_1 - P_2)}{2KL}\right) r = \alpha r$$

where $\alpha = \dfrac{(P_1 - P_2)}{2KL}$

Hence

$$\frac{dv}{dr} = -\alpha^{1/n} r^{1/n}$$

Integration gives

$$V = -\alpha^{1/n} \frac{r^{(1/n)+1}}{\dfrac{1}{n}+1} + C$$

since $V = 0$ for $r = R$, then

$$C = \alpha^{1/n} \left(\frac{n}{n+1}\right) R^{(n+1)/n}$$

and

$$V = \left[\frac{(P_1 - P_2)}{2KL}\right]^{1/n} \left(\frac{n}{n+1}\right) R^{(n+1)/n} \left[1 - \left(\frac{r}{R}\right)^{(n+1)/n}\right]$$

The volumetric flowrate through the tube

$$Q = \int_0^R 2\pi r V dr$$

$$= \left[\frac{(P_1 - P_2)}{2KL}\right]^{1/n} \left(\frac{n}{n+1}\right) R^{(n+1)/n} 2\pi \int_0^R r \left[1 - \left(\frac{r}{R}\right)^{(n+1)/n}\right] dr$$

$$= \left[\frac{(P_1 - P_2)}{2KL}\right]^{1/n} \left(\frac{n}{n+1}\right) R^{(n+1)/n} 2\pi \left[\frac{R^2}{2} - \left(\frac{n}{3n+1}\right) \frac{R^{(3n+1)/n}}{R^{(n+1)/n}}\right]$$

$$= \left[\frac{(P_1 - P_2)}{2KL}\right]^{1/n} \left(\frac{n}{n+1}\right) R^{(3n+1)/n} 2\pi \left[\frac{n+1}{2(3n+1)}\right]$$

$$= \left[\frac{(P_1 - P_2)}{2KL}\right]^{1/n} \frac{\pi n}{(3n+1)} R^{(3n+1)/n}$$

when $n = 1$, $\tau = -K(du/dr)$, $K = \mu$ i.e. the liquid is Newtonian.

Therefore

$$Q = \left[\frac{P_1 - P_2}{2\mu L} \right] \frac{\pi}{4} R^4 = \frac{\pi(P_1 - P_2)R^4}{8\mu L}$$

which is the familiar Hagen/Poiseuille equation.

Note: when $n < 1$ the liquid is said to be pseudoplastic

$n = 1$ Newtonian liquid

$n > 1$ the liquid is said to be dilatant.

Example 3-4: Show that the flowrate/pressure drop characteristic for the flow of a viscous liquid through a tube can be predicted from the rheogram for that liquid, using the expression

$$Q = -\frac{\pi R^3}{\tau_w^3} \int_0^{\tau_w} \tau^2 f(\tau) d\tau$$

where Q is the volumetric flowrate through a tube of radius R when the shear stress at the tube wall is τ_w. Show that this expression leads to the familiar Hagen–Poiseuille equation when applied to a Newtonian liquid.

Solution A section of the tube is shown in Fig. 3-5(a). Assuming that the flow is fully developed, and that steady conditions prevail,

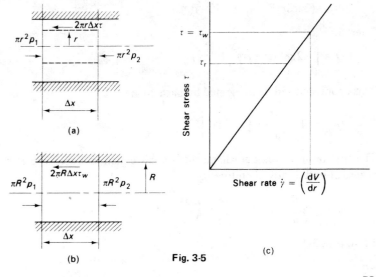

Fig. 3-5

59

then the forces acting on a cylindrical element of the liquid will be in equilibrium, and therefore the net force on the element will be zero; therefore

$$\pi r^2 p_1 = \pi r^2 p_2 + 2\pi r \Delta x \tau$$

$$r\Delta P = 2\Delta x \tau \qquad \Delta P = (p_1 - p_2)$$

That is

$$\tau = \frac{r}{2} \frac{\Delta P}{\Delta x} \qquad \text{(see Fig. 3-5(a))}$$

Similarly

$$\tau_w = \frac{R}{2} \frac{\Delta P}{\Delta x} \qquad \text{(see Fig. 3-5(b))} \tag{3-4}$$

Hence

$$\frac{\tau}{\tau_w} = \frac{r}{R} \tag{3-5}$$

Eqn 3-5 is applicable irrespective of the nature of the flow, that is to say whether it is laminar or turbulent. It shows the shear stress in the liquid varies linearly along a tube radius, being zero at the centreline and a maximum (τ_w) at the tube wall. The volumetric flowrate through an annular element of the tube cross section is given by

$$dQ = (2\pi r dr)V \qquad V = f(r)$$

Therefore

$$Q = \int_0^R 2\pi r V dr = \pi \int_0^R V d(r^2)$$

This expression can be integrated by parts to give

$$Q = \pi \left[Vr^2 \Big|_0^R - \int_0^R r^2 dV \right]$$

The first term on the right hand side of this equation is zero since Vr^2 is zero at both limits, therefore

$$Q = -\pi \int_0^R r^2 dV = -\pi \int_0^R r^2 \left(\frac{dV}{dr} \right) dr$$

From eqn. 3-5

$$r^2 = \frac{R^2}{\tau_w^2}\, \tau^2 \quad \text{and} \quad dr = \frac{R}{\tau_w}\, d\tau$$

Therefore

$$Q = -\pi \int_0^R \frac{R^3}{\tau_w^3}\, \tau^2 \left(\frac{dV}{dr}\right) d\tau$$

Given the rheogram of a particular liquid we have

$$\tau = f\left(\frac{dV}{dr}\right)$$

Therefore

$$\frac{dV}{dr} = f'(\tau) \quad \text{and} \quad Q = \frac{-\pi R^3}{\tau_w^3} \int_0^{\tau_w} \tau^2 f'(\tau)\, d\tau \qquad (3\text{-}6)$$

For the Newtonian liquid we have the rheogram shown in Fig. 3-5(c)

$$\tau = -\mu \frac{dV}{dr} = -\mu \dot{\gamma} \quad \text{and} \quad \frac{dV}{dr} = -\frac{\tau}{\mu}$$

Therefore

$$Q = \frac{-\pi R^3}{\tau_w^3} \int_0^{\tau_w} \tau^2 \left(-\frac{\tau}{\mu}\right) d\tau = \frac{\pi R^3}{\tau_w^3} \left(\frac{\tau^4}{4\mu}\right)^{\tau_w} = \frac{\pi R^3 \tau_w}{4\mu}$$

From eqn. 3-4

$$\tau_w = \frac{R}{2} \frac{\Delta P}{\Delta x}$$

Therefore

$$Q = \frac{\pi R^4}{8\mu} \frac{\Delta P}{\Delta x} \quad \text{(Hagen–Poiseuille equation)}$$

In the general case, the evaluation of Q for a range of values of τ_w (or $\Delta P/\Delta x$) can be carried out by numerical integration of eqn. 3-6.

3.2.2 Pressure drop calculations

In Examples 3-2, 3-3 and 3-4, the method of calculating flow rate/pressure drop characteristics from first principles was demonstrated. The equations which were derived in these examples are valid

61

only provided laminar flow conditions prevail, that is to say that the shear stress in the liquid in question is described by equations such as 3-1 to 3-3. When turbulent conditions exist the actual shear stress in the liquid is greater than would be predicted by these equations, and therefore the resultant equations for flowrate and pressure drop are no longer applicable. The transition from laminar to turbulent flow in tube flow is known to be associated with the Reynolds number ($DV\rho/\mu$). In general, if this parameter does not exceed about 2000, then for a Newtonian liquid, flow conditions will be laminar, shear stress will be given by Newton's law and the Hagen–Poiseuille equation will apply.

At present, a detailed description of turbulent flow is not generally available, and therefore design equations for these conditions are based on empirical equations. Dimensional analysis is helpful in determining convenient forms of parameter for such equations. For example, if the shear stress at the wall of a tube τ_w is assumed to be a function of tube diameter D, the fluid density ρ, viscosity μ, and the average velocity of flow V, then

$$\tau_w = \phi(D, \rho, \mu, V)$$

It is left to the reader to show that dimensional analysis (i.e. the method of indices, or the Buckingham π theorem) indicates that Eqn. 3-7 can be rewritten

$$\frac{\tau_w}{\rho V^2} = \phi\,(DV\rho/\mu) = \phi(Re) \tag{3-8}$$

or

$$\frac{\tau_w}{\tfrac{1}{2}\rho V^2} = \phi'\,(Re) \tag{3-9}$$

($DV\rho/\mu$) is the familiar Reynolds number Re, and ($\tau_w/\tfrac{1}{2}\rho V^2$) is one form of a parameter called the friction factor f. If a series of experiments is conducted in which f is determined for a wide range of Reynolds number then the characteristic shown in Fig. 3-6 results.

In the laminar region, for $Re \leqslant 2000$

$$f = \frac{16}{Re} \text{ , therefore}$$

$$\frac{\tau_w}{\tfrac{1}{2}\rho V^2} = \frac{16\mu}{DV\rho}$$

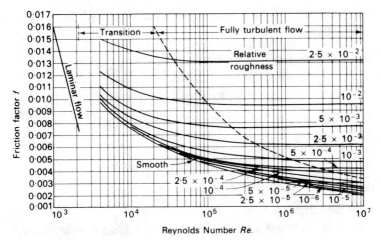

Fig. 3-6

But

$$\tau_w = \frac{R}{2}\frac{\Delta P}{\Delta x} = \frac{4\mu V}{R}$$

and so

$$V = \frac{R^2}{8\mu}\frac{\Delta P}{\Delta x}$$

or since $V = Q/\pi R^2$

$$Q = \frac{\pi R^4}{8\mu}\frac{\Delta P}{\Delta x} \quad \text{(Hagen–Poiseuille equation)}$$

In the region $2000 < Re < 5000$, the flow is unsteady and meaningful results cannot be obtained. For $Re > 5000$, the roughness of the tube surface is an important parameter and is introduced by means of a term called the relative roughness, which is the dimensionless ratio of the mean height of surface roughness, to the tube diameter D.

Example 3-5: Oil is to be pumped to an elevated storage tank through a pipeline 40 mm inside diameter, 150 m long. If the pipeline discharges into the tank at a point 10 m above the pump outlet, what pressure is required at the pump outlet in order to maintain an oil flowrate of 0·9 kg/s? The density and viscosity of the oil is 882 kg/m³

63

and $3\cdot6 \times 10^{-3}$ Ns/m^2 respectively. The pipe wall has a relative roughness of 10^{-4}.

Solution   Average flow velocity V of oil

$$= (0\cdot9/882)/[\pi(40/1000)^2/4] = 0\cdot81 \text{ m/s}$$
$$R_e = (DV\rho/\mu) = (40/1000)\, 0\cdot8 \times 882/3\cdot6 \times 10^{-3} = 7938$$

From Fig. 3-6, with $R_e = 7938$, relative roughness $= 10^{-4}$,

$f = 0\cdot008$.

Since $f = \tau_w/(\tfrac{1}{2}\rho V^2) = [(R/2)\Delta p/\Delta x]/(\tfrac{1}{2}\rho V^2)$

then

$$\Delta P = 2f(L/D)\rho V^2$$
$$= 2 \times 0\cdot008 \times (150 \times 1000/40)882 \times (0\cdot81)^2 = 34\cdot7 \text{ kN/m}^2$$

(3-10)

Static head required

$$= H'\rho g = 10 \times 882 \times 9\cdot81 = 86\cdot5 \text{ kN/m}^2$$

Total pressure required at pump outlet = frictional pressure drop
+ static pressure head

$$= 121\cdot2 \text{ kN/m}^2$$

3.3 Drag forces on particles

3.3.1 Terminal velocity calculations

Example 3-6:   What is meant by the terminal velocity of particles settling in a liquid or gaseous medium? Calculate the terminal velocity of spherical particles, $0\cdot5$ mm diameter, settling in a liquid having a density and viscosity of 10^3 kg/m^3 and $0\cdot1$ Ns/m^2 respectively. The density of the solid is $4\cdot3 \times 10^3$ kg/m^3. Assume that the drag force on the particles is given by the Stoke's equation $F = 3\pi\mu DV$.

Solution   When a particle is released from rest in a stationary fluid it accelerates initially under the action of a force equal to the difference between its weight and the bouyancy force. However, once motion commences, this net force, and therefore particle acceleration, decreases owing to the drag force on the particle resulting from the relative motion between the particle and the fluid. This drag force increases with particle velocity. At a particular value of particle velocity, the sum of the drag and bouyancy force equals the weight of the particle and the particle acceleration is zero. The particle then continues to fall with constant velocity V_t. This velocity is called the terminal velocity of the particle.

Weight of particle diameter D is $\dfrac{\pi D^3}{6} \rho_s g$

Bouyancy force is $\dfrac{\pi D^3}{6} \rho g$

where ρ_s and ρ are the solid and liquid densities respectively.

Drag force is $3\pi \mu D V$

If V_t is the terminal velocity of the particle, then

$$\frac{\pi D^3}{6}(\rho_s - \rho)g = 3\pi \mu D V_t$$

Hence

$$V_t = \frac{D^2 g}{18\mu}(\rho_s - \rho) = \frac{(0 \cdot 5 \times 10^{-3})^2 \; 9 \cdot 81 \; (4 \cdot 3 - 1)10^3}{18 \times 0 \cdot 1}$$

$$= 4 \cdot 49 \times 10^{-3} \text{ m/s} \equiv 4 \cdot 5 \text{ mm/s}$$

Note: The drag equation $F = 3\pi \mu D V$ should not be applied for

$$Re' \left(= \frac{DV}{\mu} \rho \right) > 0 \cdot 2$$

In the present example

$$Re' = 0 \cdot 5 \times 10^{-3}(4 \cdot 49 \times 10^{-3}) \; 10^3 / 0 \cdot 1 = 0 \cdot 022$$

Example 3-7: Assuming that a graphical plot of particle drag coefficient against particle Reynolds number for spherical particles is available, show that these data may be used to calculate

(a) terminal velocity; given particle diameter D, particle density ρ_s liquid density ρ and viscosity μ, and

(b) particle diameter; given terminal velocity V_t, particle density ρ_s liquid density ρ and viscosity μ,

without recourse to trial and error solutions. Indicate how this technique is used to calculate the terminal velocity of the particles in Example 3-5.

Solution The drag coefficient C_D of a particle is defined by the equation

$$C_D = (\text{drag force } F/\text{unit projected area})/\tfrac{1}{2}\rho V^2$$

and the particle Reynolds number $R_e = (DV\rho/\mu)$
Figure 3-7 shows the $C_D:R_e{}'$ characteristic for spherical particles. For

$$R_e{}' < 0{\cdot}2 \quad C_D = 24/R_e{}' = [F/(\pi D^2/4)]/\tfrac{1}{2}\rho V^2$$

from which $f = 3\pi\mu DV$ the familiar Stoke's Law.

For $R_e{}' < 800$ the $C_D:R_e{}'$ characteristic can be approximated by the equation

$$C_D = (24/R_e{}')(1 + 0{\cdot}15(R_e{}')^{0{\cdot}687})$$

Assuming a particle has reached its terminal velocity, then

$$F = (\pi D^3/6)g(\rho_s - \rho)$$

Therefore

$$C_D(R_e{}')^2 = 4D^3\rho g(\rho_s - \rho)/3\mu^2$$

and

$$C_D/R_e{}' = 4g\mu(\rho_s - \rho)/3\rho^2 V t^3$$

The original C_D, $R_e{}'$ characteristic of Fig. 3-7, (or its approximating function), can now be used to produce two additional characteristics

(i) $C_D(R_e{}')^2$ against $R_e{}'$, and
(ii) $C_D/R_e{}'$ against R_e.

For a problem where terminal velocity is required, given D, ρ_s and ρ as in Example 3-6, calculate

$$C_D(Re')^2 = 4(0{\cdot}5 \times 10^{-3})^3 10^3(4{\cdot}3 - 1)10^3 \times 9{\cdot}81/3(0{\cdot}1)^2 = 0{\cdot}54$$

Using (i)

$$C_D(Re')^2 = 0{\cdot}54 \qquad Re' = 0{\cdot}023$$

Therefore

$$0{\cdot}023 = ((0{\cdot}5 \times 10^{-3}) \times V_t \times 10^3)/0{\cdot}1$$

and

$$V_t = 4{\cdot}6 \times 10^{-3} \text{ m/s}$$

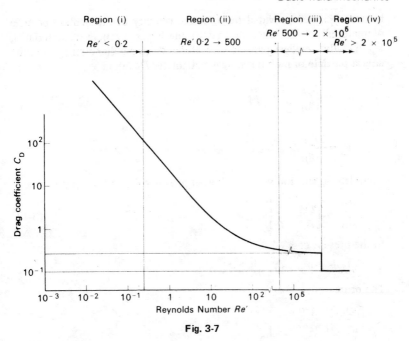

Region (i)   Region (ii)   Region (iii)   Region (iv)

Re' 500 → 2 × 10^5

$Re' < 0.2$ Re' 0.2 → 500 $Re' > 2 × 10^5$

Fig. 3-7

3.3.2 Particle/fluid separation

Example 3-8: A mixture of two materials A and B is to be separated into fractions of materials A and B only by making use of the different terminal velocities of the particles. If the density of material A, material B, and the suspending liquid are in the ratios 1·9 and 1·35 to 1 respectively, show that for complete separation the ratio of the maximum diameter of particles of material B to the minimum diameter of particles of material A must not exceed about 1·6.

The particles may be assumed to be spherical, the drag force being described by the Stokes Law ($F = 3\pi\mu DV$).

Solution: The terminal velocity of spherical particles V_t was shown in Example 3.5 to be given by the equation

$$V_t = \frac{D^2 g}{18\mu} [\rho_s - \rho]$$

It can be seen from this equation that terminal velocity increases with particle material density and particle diameter; complete separation will be accomplished provided the terminal velocity of the smallest particle of material A, V_A, exceeds that of the largest particle of material B, V_B. Let the smallest particle of material A have a diameter d_A and the largest particle of material B have a diameter D_B, then

$$V_A = \frac{d_A^2 g}{18\mu} [\rho_A - \rho]$$

and

$$V_B = \frac{D_B^2 g}{18\mu} [\rho_B - \rho]$$

Complete separation will be achieved provided $V_A > V_B$, that is

$$\frac{D_B}{d_A} < \left[\frac{\rho_A - \rho}{\rho_B - \rho}\right]^{0.5} \tag{3-11}$$

In the present case

$$\rho_A = 1 \cdot 9 \, \rho \text{ and } \rho_B = 1 : 35 \, \rho$$

Therefore

$$\left[\frac{D_B}{d_A}\right]\left[\frac{1 \cdot 9 - 1}{1 \cdot 35 - 1}\right]^{0.5} = 1 \cdot 61$$

If the particle size range in the original mixture was such that the above ratio were, say 2, then complete separation into fractions of materials A and B only, will not be possible with the liquid specified. In order to achieve complete separation a liquid with density $\rho = x\rho_A$ must be used, where

$$\frac{D_B}{d_A} = 2 = \left[\frac{\rho_A - \rho}{\rho_B - \rho}\right]^{0.5}$$

that is

$$4 = \frac{\rho_A - \rho_f}{\rho_B - \rho_f} = \left[\frac{\rho_A - x\rho_A}{\dfrac{1 \cdot 35}{1 \cdot 9} \rho_A - x\rho_A}\right]$$

Since

$$\rho_A = 1 \cdot 9\rho \text{ and } \rho_B = 1 \cdot 35 \, \rho$$

the equation gives

$$x = 0 \cdot 613$$

Therefore, ρ must not be $< 0 \cdot 613 \, \rho_A$ for this mixture.

3.3.3 Particle/gas separation in cyclones

Example 3-9: Tests on a wide range of cyclones with geometry as shown generally in Fig. 3-8 indicate that the following assumptions can be made about their performance:

1. Radial gas velocity is a function of radius only.
2. Tangential velocities are inversely proportional to the square root of the radius.
3. Tangential velocity at the cyclone wall is equal to inlet duct gas velocity.
4. Radius of central turbulent core is equal to approximately 20% outlet duct diameter.

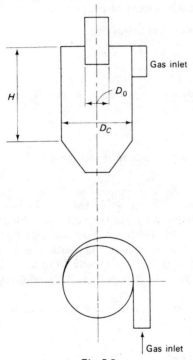

Fig. 3-8

Use these assumptions to estimate the minimum particle diameter which would be retained by the cyclone.

An existing cyclone is to be used to separate solid particles from a CO_2 stream at 300 K. The cyclone, which has an overall diameter three times the gas outlet diameter, has a depth of 1·3 m. The inlet duct is

200 mm wide by 350 mm deep. If the gas flowrate is to be 0·5 kg/s, and it contains a minimum particle size of 15×10^{-6} m (15 μm), would the cyclone be effective? Additional data:—

CO_2 density: 1·79 kg/m^3
CO_2 viscosity: $1·49 \times 10^{-5}$ Ns/m^2
Solid density: 1900 kg/m^3

Solution Consider a particle diameter d, density ρ_s at a radius r. Let the tangential and radial velocity at radius r be v_t and v_r respectively, then

$$v_t = r\omega$$

where ω is the angular velocity of the particle.

Centrifugal acceleration at radius r is

$$r\omega^2 = \frac{v_t^2}{r}$$

Centrifugal force on the particle is

$$\frac{\pi d^3}{6} \rho_s \frac{v_t^2}{r}$$

Radial drag force on the particle at radius r is

$$3\pi\mu dv_r$$

In the cyclone, particles are subjected to two opposing forces, the centrifugal force induced by the tangential velocity of the gas and the drag force induced by radial gas flow towards the gas outlet. Both forces are functions of radius and particle size as shown above, therefore particles of different size tend to rotate at different radii. In good cyclone design, v_t is as high as possible, whilst v_r is made as small as possible.

The radius at which a particle will tend to rotate can be found by equating the radial forces

$$\frac{\pi d^3}{6} \rho_s \frac{v_t^2}{r} = 3\pi\mu dv_r$$

Therefore

$$d^2 = \frac{18 \, r \, \mu v_r}{\rho_s v_t^2}$$

Thus, the smallest particles are located near to the centre of the cyclone. Assuming that particles which enter the turbulent cone are lost from the cyclone then d_m is given by

$$d_m^2 = \frac{18 \times 0 \cdot 2 D_0 \mu v_r}{\rho_s v_t^2}$$

for $r = 0 \cdot 2 D_0$ (from assumption 4)

If the mass flowrate of gas through the cyclone is G kg/s then

$$v_r = \frac{G}{2\pi r H \rho} = \frac{G}{2\pi 0 \cdot 2 D_0 H \rho} \quad \text{for } r = 0 \cdot 2 D_0$$

If the tangential velocity of gas at the outer wall of the cyclone is v_{t0} then assumption 2 gives

$$v_t = v_{t0} (D_c/2r)^{\frac{1}{2}} = v_{t0} (D_c/2 \times 0 \cdot 2 D_0)^{\frac{1}{2}} = \frac{G}{A_i \rho} (D_c/2 \times 0 \cdot 2 D_0)^{\frac{1}{2}}$$

since

$$v_{t0} = G/A_i \rho \text{ inlet gas velocity, (assumption 3)}$$

Hence

$$d_m^2 = \frac{3 \cdot 6 \mu D_0 \rho A_i^2}{\pi H \rho_s G D_c}$$

In the present case

$$\frac{D_0}{D_c} = \frac{1}{3}$$

Therefore

$$d_m^2 = \frac{3 \cdot 6 \mu \rho A_i^2}{3\pi H \rho_s G} = \frac{0 \cdot 38 \mu \rho A_i^2}{H \rho_s G}$$

Hence

$$d_m^2 = \frac{0 \cdot 38 (1 \cdot 49 \times 10^{-5}) 1 \cdot 79 (0 \cdot 07)^2}{1 \cdot 3 \times 1900 \times 0 \cdot 5} = 40 \cdot 2 \times 10^{-12}$$

and

$$d_m = 6 \cdot 35 \times 10^{-6} \text{ m } (6 \cdot 35 \ \mu\text{m})$$

This result indicates that the cyclone would be effective for the proposed duty since the minimum particle size in the gas is 15 μm. In practice, the cyclone becomes inefficient when the gas stream contains a large proportion of particles with diameter less than about 10 μm and therefore the calculated value for minimum size retained must be treated with caution. Problems also exist with the larger particles which may bounce off the wall of the cyclone to be entrained in the outgoing gas. For effective cyclone operation larger-sized particles should be removed prior to the cyclone. The effluent gas will then contain essentially fine material which can be removed by scrubbing, filtration or electrostatic precipitation, if this is necessary.

3.3.4 Particle/liquid separation in centrifugal classifiers

Example 3-9:   A dilute slurry is fed onto the inner surface of a vertical cylinder of radius R, rotating at N Hz. Under normal operating conditions the radius of the liquid surface is R_i. If the slurry contains particles of diameter d, density ρ_s, and the liquid density and viscosity is ρ and μ respectively, show that the time taken for a particle to move from the liquid surface to the cylinder wall is given by

$$t = \frac{18\mu}{d^2(\rho_s - \rho)\omega^2} \ln \frac{R}{R_i}$$

It may be assumed that the particles experience a drag force given by Stoke's Law ($F = 3\pi\mu dv$).

Solution: A particle rotating in the liquid at radius r ($R_i < r < R$) will be subjected to a centrifugal acceleration given by $r\omega^2$, where ω is the angular velocity of the particle (= $2\pi N$ 1/s).

As a result of this acceleration the particle will experience a force given by its mass multiplied by the acceleration, i.e. $(\pi d^3/6)\rho_s r\omega^2$. If the actual particle acceleration is denoted by d^2r/dt^2 then

$$\underbrace{\frac{\pi d^3}{6}\rho_s \frac{d^2 r}{dt^2}}_{\substack{\text{mass x} \\ \text{acceleration}}} = \underbrace{\frac{\pi d^3}{6}(\rho_s - \rho)r\omega^2 - 3\pi\mu d \frac{dr}{dt}}_{\text{(net force on the particle)}}$$

In this equation, the drag force on the particle has been assumed to be that described by Stoke's Law. In this situation, the velocity of a particle does not reach a terminal value since the force on a particle increases with increasing radius. The above equation can be solved either by a numerical or graphical technique; however, if the acceleration term is neglected, the equation reduces to

$$\frac{dr}{dt} = \frac{d^2(\rho_s - \rho)r\omega^2}{18\,\mu}$$

or

$$\frac{18\,\mu}{d^2(\rho_s - \rho)\omega^2}\,\frac{dr}{r} = dt \quad \text{(by separation of variables)}$$

Integration gives

$$\frac{18\,\mu}{d^2(\rho_s - \rho)\omega^2}\,\ln r = t + c$$

Boundary conditions

$$r = R_i \quad t = 0$$
$$r = R \quad t = t$$

Therefore

$$t = \frac{18\,\mu}{d^2(\rho_s - \rho)\omega^2}\,\ln (R/R_i)$$

If the slurry is fed to this unit at a volumetric flowrate of Q, and the effective length of the cylinder is L then the approximate residence time for an element of slurry in the cylinder

$$t = \frac{\text{distance}}{\text{velocity}} = \frac{L}{Q/\pi(R^2 - R_i^2)} = \frac{L\pi(R^2 - R_i^2)}{Q} = \frac{\text{volumetric hold up}}{\text{volumetric flowrate}}$$

If the smallest particle retained in the unit specified above has a diameter d_m, its residence time is given by

$$t = \frac{18\,\mu}{d_m^2(\rho_s - \rho)\omega^2}\,\ln (R/R_i) = \frac{L\pi(R^2 - R_i^2)}{Q}$$

Therefore

$$d_m = \left[\frac{18\,\mu\,Q\,\ln (R/R_i)}{L\pi(R^2 - R_i^2)(\rho_s - \rho)\omega^2}\right]^{\frac{1}{2}}$$

A discussion of the centrifugal classifier, including the practical aspects of its construction can be found in Coulson and Richardson which also provides further references to the more detailed technical literature on this subject.

3.4 Flow in packed beds

The calculation of pressure drop in tubes is conveniently described by equations relating friction factor and Reynolds number. This procedure

73

has been outlined in section 3.2.2. A similar procedure is adopted for the calculation of pressure drop accompanying the flow of a fluid through a packed bed. In this case, a modified friction factor f' and Reynolds number Re' are defined as

$$Re' = \frac{\rho V_s}{s\mu(1-e)} \quad \text{and} \quad f' = \frac{\Delta P e^3}{Ls(1-e)\rho V_s^2}$$

where V_s (the superficial velocity of flow)

$$= \frac{\text{volumetric flowrate through the bed}}{\text{cross sectional area of bed normal to flow direction}}$$

Example 3-10: A filter bed has the following characteristics: bed diameter (D) 2 m, bed depth (L) 0·6 m, bed porosity (e) 0·48 and specific surface (S) 15 500 m^2/m^3. If a process liquid has a density and viscosity of 10^3 kg/m^3 and 10^{-3} Ns/m^2 respectively, calculate the mass flowrate (kg/s) through the bed if the pressure drop across it is maintained at 10^4 N/m^2.

Solution   It may be assumed that $f' = 5/Re'$ provided $Re' < 2$

then

$$f' = \frac{\Delta P e^3}{Ls(1-e)\rho V_s^2} = \frac{5s\,\mu(1-e)}{\rho V_s}$$

Hence

$$V_s = \frac{e^3 \Delta P}{5s^2\,\mu(1-e)^2 L} \quad \text{This is the Carmen/Kozeny equation.}$$

Let

$$r = \frac{5s^2(1-e)^2}{e^3}, \text{ then}$$

$$V_s = \frac{\Delta P}{r\mu L}$$

r is called the specific resistance of the bed. In the present problem

$$r = \frac{5(15\,500)^2\,(1-0\cdot48)^2}{(0\cdot48)^3} = 29\cdot4 \times 10^8 \text{ m}^{-2}$$

therefore

$$V_s = \frac{\Delta P}{r\mu L} = \frac{10^4}{29 \cdot 4 \times 10^8 \times 10^{-3} \times 0 \cdot 6}$$
$$= 0 \cdot 0057 \text{ m/s}$$

Mass flowrate through bed is

$$\frac{\pi D^2}{4} \times 0 \cdot 0057 \rho = \frac{\pi 2^2}{4} \times 0 \cdot 0057 \times 10^3 = 17 \cdot 9 \text{ kg/s}$$

Check on Re':

$$Re' = \frac{\rho V_s}{s\mu(1 - e)} = \frac{10^3 \times 0 \cdot 0057}{15\,500 \times 10^{-3}\,(1 - 0 \cdot 48)}$$
$$= 0 \cdot 71$$

3.4.1 Filtration

The equation governing flow in packed beds of particles, $V_s = \Delta P/r\mu L$, can be used to calculate rates of filtration under certain conditions.

If the instantaneous volumetric flowrate of filtrate from the bed is $\dfrac{dQ}{dt}$, and the bed area is A, then

$$V_s = \frac{1}{A}\frac{dQ}{dt} = \frac{\Delta P}{r\mu L}$$

or

$$\frac{dQ}{dt} = \frac{A\Delta P}{r\mu L}$$

Let the volume of cake deposited with the passage of unit volume of filtrate be q, then for deposition of a cake of area A, depth L, the volume of filtrate passed Q, is given by

$$Qq = AL$$

therefore

$$\frac{dQ}{dt} = \frac{A^2 \Delta P}{r\mu q Q}$$

This is the general filtration equation.

Example 3-12: A test filtration is carried out on a process slurry under constant pressure conditions. The area of the laboratory filter is $0.0465 \, \text{m}^2$ and the pressure drop across the filter is maintained at $68 \, \text{kN/m}^2$. Under these conditions, $0.25 \times 10^{-3} \, \text{m}^3$ of filtrate are passed in 300 s whilst $0.4 \times 10^{-3} \, \text{m}^3$ are passed in 600 s.

Under process conditions, the filtration area will be $1.12 \, \text{m}^2$. Filtration will be carried out initially under constant flowrate conditions until the pressure drop across the filter reaches $415 \, \text{kN/m}^2$. This operation will take 180 s. Filtration will then continue for a further 900 s. The filter cake will then be washed for a further 600 s. Calculate the total volume of filtrate passed during this operation, and also the volume of wash water used. The cake may be assumed to be incompressible. The filter medium will be common to both test and process operations.

Solution   The filter equation

$$\frac{dQ}{dt} = \frac{A^2 \Delta P}{r\mu q Q}$$

does not account for the resistance to flow presented by the filter medium on which the cake is to be built up. If this resistance is equivalent to an additional layer of cake l thick then

$$\frac{dQ}{dt} = \frac{A\,\Delta P}{r\mu(L + l)} = \frac{A\,\Delta P}{r\mu q \left(\dfrac{Q}{A} + \dfrac{l}{q} \right)} \tag{3-12}$$

since

$$Qq = AL$$

since the laboratory filtration is carried out at constant pressure, eqn. 3-12 can be integrated directly. The cake is assumed to be incompressible (e and s constant) therefore the specific resistance r of the cake will be constant. From eqn. 3-12 we have

$$Q^2 + 2A \frac{l}{q} Q = \frac{2A^2 \Delta P}{r\mu q} t$$

Therefore

$$(0.25 \times 10^{-3})^2 + 2 \times 0.0465 \times (0.25 \times 10^{-3}) \frac{l}{q}$$

$$= \frac{2(0.0465)^2 \times 68 \times 10^3 \times 300}{r\mu q}$$

and

$$(0 \cdot 4 \times 10^{-3})^2 + 2 \times 0 \cdot 0465 \times (0 \cdot 4 \times 10^{-3})\frac{l}{q}$$

$$= \frac{2(0 \cdot 0465)^2 \times 68 \times 10^3 \times 600}{r\mu q}$$

which can be solved to give

$$r\mu q = 49 \cdot 13 \times 10^{10} \qquad \frac{l}{q} = 4 \cdot 19 \times 10^{-3}$$

Process filtration: area $A = 1 \cdot 12$ m^2, pressure $\Delta P = 415 \times 10^3$ N/m^2, constant rate period $t_1 = 180$s, Q_1 required. Now

$$\frac{dQ}{dt} = \text{constant} = \frac{Q_1}{t_1} = \frac{A \Delta P}{r\mu q \left[\dfrac{Q_1}{A} + \dfrac{l}{q}\right]}$$

Therefore

$$Q_1^2 + A \frac{l}{q} Q_1 = \frac{\Delta P A^2 t_1}{r\mu q}$$

so

$$Q_1^2 + 1 \cdot 12 \times 4 \cdot 19 \times 10^{-3} Q_1 = \frac{415 \times 10^3 (1 \cdot 12)^2 \, 180}{49 \cdot 13 \times 10^{10}}$$

For the constant pressure period

$$\frac{dQ}{dt} = \frac{A \Delta P}{r\mu q \left[\dfrac{Q}{A} + \dfrac{l}{q}\right]}$$

$$Q dQ + A \frac{l}{q} dQ = \frac{A^2 \Delta P}{r\mu q} dt$$

$$Q^2 + 2A \frac{l}{q} Q = \frac{A^2 \Delta P}{r\mu q} t$$

$$(Q_2^2 - Q_1^2) + 2A - (Q_2 - Q_1) = \frac{A^2 \Delta P}{r\mu q}(t_2 - t_1)$$

Therefore

$$Q_1 = 11 \cdot 7 \times 10^{-3} \text{m}^3 \quad t_2 - t_2 = 900 \text{ s}$$

Substitution and solution gives

$$Q_2 = 41 \cdot 95 \times 10^{-3} \text{ m}^3$$

The rate of filtrate flow at the end of the filtration process

$$\frac{dQ}{dt} = \frac{A \Delta P}{r \mu q \left[\dfrac{Q}{A} + \dfrac{l}{q} \right]}$$

with $Q = Q_2 = 41 \cdot 95 \times 10^{-3} \text{ m}^3$

$$\frac{dQ}{dt} = 0 \cdot 0229 \times 10^{-3} \text{ m}^3/\text{s}$$

If the wash water has the same properties as the filtrate the quantity required

$$Q_3 = \frac{dQ}{dt} \times \text{washing time} = 0 \cdot 229 \times 10^{-3} \times 600 = 13 \cdot 7 \times 10^{-3} \text{ m}^3$$

List of symbols

A	area	(m^2)
c	integration constant	$(-)$
d, D	diameter	(m)
e	voidage (or porosity)	$(-)$
$f(\,)$, $f'(\,)$	functions of parameters in parenthesis	
g	gravitational acceleration	(m/s^2)
G	mass flowrate	(kg/s)
h	semi-height of duct	(m)
H	effective height of cyclone	(m)
K	consistency index of power law fluid	(Ns^n/m^2)
l	equivalent thickness of filter medium	(m)
L	length of duct, depth of packed or filter bed	(m)
n	flow behaviour index of power law fluid	$(-)$
N	rotary speed	(Hz)
p	pressure	(N/m^2)
$\Delta P / \Delta x$	pressure gradient	$((\text{N/m}^2)/\text{m})$
Q	volume of filtrate passed	(m^3)
r	specific resistance of filter cake	(m^{-2})
r, R	radius	(m)

s	specific surface	(m^2/m^3) or $(1/m)$
t	time	(s)
T	torque	(Nm)
V, v	velocity	(m/s)
W	duct width	(m)
$\dot{\gamma}$	shear rate	(1/s)
μ	viscosity	(Ns/m^2)
ρ, ρ_f	liquid density	(kg/m^3)
ρ_s	solid density	(kg/m^3)
τ	shear stress	(N/m^2)
τ_y	yield stress	(N/m^2)
τ_w	shear stress at wall	(N/m^2)
ϕ	angle	$(-)$
ω	angular velocity	(1/s)

Dimensionless numbers

c_D	drag coefficient:– (force/unit projected area)/½ ρV^2
f	friction factor $(\tau_w/½ \rho V^2)$
Re	Reynolds number $(DV\rho/\mu)$
Re'	modified Reynolds numbers for particles or packed beds, as defined in the text.

Subscripts

A	referring to material A
B	referring to material B
1	referring to upstream conditions
2	referring to downstream conditions.

Further reading

COULSON, J. M. and RICHARDSON, J. F. *"Chemical Engineering"* Volumes I and II 2nd Editions. Pergamon Press, Oxford (1969).

BIRD, R. B., STEWART, W. F. and LIGHTFOOT, E. N. *"Transport Phenomena"* Wiley, New York, (1960).

STREETER, V. L. *"Fluid Mechanics"* 5th Edition McGraw-Hill, New York (1971).

BRIDGMAN, P. W. *"Dimensional Analysis"* Yale University Press, New Haven, Conn. (1931).

LANGHAAR, H. L. *"Dimensional Analysis and Theory of Models"* Wiley, New York (1951).

WILKINSON, W. L. *"Non-Newtonian Fluids"* Pergamon Press, Oxford (1960).

SKELLAND, A. H. P. *"Non-Newtonian Flow and Heat Transfer"* Wiley, London (1967).
BATCHELOR, G. K. *"An Introduction to Fluid Dynamics"* Cambridge University Press (1970).
FRANCIS, J. R. D. *"A Textbook of Fluid Mechanics"* (for engineering students) S.I. Units Edward Arnold, London (1971).
HOLLAND, F. A. *"Fluid Flow for Chemical Engineers"* Edward Arnold, London (1973).

Problem 3-1: A process liquid fills the space between two long co-axial cylinders which form part of a device for measuring the viscosity of a liquid. The outer cylinder has an internal diameter d_2 of 150 mm and is fixed. The inner cylinder, which is solid has a diameter d_1 of 145 mm, a length of 10 mm and is free to rotate. In order to rotate the inner cylinder at a steady 30 revolutions/min it is found that a torque of 9×10^{-2} Nm must be applied to the cylinder. Derive an expression for the viscosity μ of the process liquid in terms of d_1, d_2, the length of the cylinder L, the torque T and the rotary speed N r.p.m. of the inner cylinder. Calculate the viscosity of the process liquid. ($\omega = 2\pi n$ 1/s)

$$\left(\mu = \frac{2T}{\pi \omega L d_1^2} \left[\frac{d_2}{d_1} - 1 \right] \quad \mu = 0.299 \text{ Ns/m}^2 \right)$$

Problem 3-2: Detailed rheological measurements show that the viscosity μ of a liquid depends upon the shear rate according to the following equation

$$\mu = \mu_0 \frac{du}{dy}$$

Show that when this liquid flows through a long tube of radius R the volumetric flowrate Q and the pressure gradient $\Delta P/\Delta L$ are related by the equation

$$Q = \left(\frac{2\pi}{7} \right) R^{7/2} \left(\frac{\Delta P/\Delta L}{2\mu_0} \right)^{1/2}$$

Hint:— Show that the liquid is a power law fluid with flow behaviour index $n = 2$.

Problem 3-3: Distinguish between 'skin friction' and 'form drag'. Illustrate your answer with reference to flow over a sphere as the free stream velocity is increased.

A slurry in which the smallest particle is 0.05 mm is to be separated into size fractions using a settling tank. The tank is 16 m long and the slurry is fed into one end of the tank at a height of 3.5 m above the base of the tank. If all the solid material is to be settled estimate the maximum flow velocity through the tank. The tank liquid is essentially water, density 10^3 kg/m^3, viscosity 10^{-3} Ns/m^2. Density of the solid: 2.1×10^3 kg/m^3.

A storage tank contains liquid and solids which must be kept in suspension by recirculation. If the minimum rise velocity in the tank is expected to be 20 mm/s calculate the maximum particle size which can be permitted in the tank. The solid density is 2.4×10^3 whilst the density and viscosity of the liquid will be 10^3 kg/m^3 and 10^{-3} Ns/m^2 respectively.

$(6.87 \times 10^{-3}$ m/s., 0.16 mm.)

Problem 3-4: Discuss the important variables which affect the performance of cyclone separators.

A cyclone separator, 200 mm diameter and 500 mm long, has an inlet duct 100 mm by 30 mm, and a gas outlet diameter 75 mm. Predict the size of the smallest particle which will be retained by the cyclone. The dust laden air enters the cyclone at the rate of 0.14 kg/s.

Density of particle material:$- 2.8 \times 10^3$ kg/m^3
Density of air :$- 1.2$ kg/m^3
Viscosity of air :$- 1.9 \times 10^{-5}$ Ns/m^2

$(17 \ \mu m)$

Problem 3-5: At one stage in a chemical process a liquid is to be passed through a bed of catalyst. The flowrate through this bed will be 45 kg/s, and it is necessary to estimate the pressure drop across this bed using data from a small prototype bed of geometrical similar particles. The test liquid will be water. The details of the process and prototype beds are given below

	Process	Prototype
Bed diameter m.	1.5	0.2
Bed depth (L) m.	1.8	0.6
Porosity (e)	0.54	0.48
Liquid density (ρ) kg/m^3	0.85×10^3	1.0×10^3
Liquid viscosity (μ) Ns/m^2	8.3×10^{-3}	1.0×10^{-3}
Specific surface (s) m^{-1}	39 300	15 500

What flowrate should be used in the prototype experiment? If the pressure drop across the prototype bed with this flowrate is 10^3 N/m^2

estimate the pressure drop across the process bed at its operating flowrate. The friction factor f' and Reynolds Number Re' for packed beds are defined as $\Delta P\, e^3/L(1 - e)s\, \rho V^2$ and $\rho V/(1 - e)s\mu$ respectively where V is the superficial velocity. It may be assumed that f' is a function of Re' only.

$$(0.043 \text{ kg/s}, 1924 \text{ kN/m}^2)$$

Problem 3-6: A slurry is supplied to a filter unit by means of a positive displacement. A by-pass line is also fitted to the pump outlet and this operates when the pressure at the filter reaches some predetermined value. In operation, this valve opens 15 minutes after the filtration begins. During this period 0.4 m^3 of filtrate is collected. Laboratory tests indicate that a satisfactory cake thickness will be achieved with the passage of a total of 1.5 m^3 of filtrate. Calculate the total filtration time. If the cake is to be washed using 0.3 m^3 of wash water at the same filter pressure, calculate the length of the washing period.

The general filtration equation $dQ/dt = C\Delta P/Q$ where C = a constant may be used *without* proof.

$$(6778 \text{ s. Washing } 2520 \text{ s})$$

Problem 3-7: A mixture of three solids A, B and C has the following composition:

Material	Density kg/m^3	Particle diameter D size range mm
A	1400	1 to 0.1
B	800	0.3 to 0.04
C	650	0.07 to 0.01

Separation of the mixture into materials A, B and C and then into size fractions is to be effected by taking advantage of the differences in terminal velocity between the particles in a liquid of density 600 kg/m^3. What degree of separation can be achieved with this liquid. If a second liquid is required calculate its density.

The particles in this mixture may be assumed spherical, and the drag coefficient for the particles should be calculated from the equation

$$C_D = \frac{F}{\frac{1}{2}\rho v^2} = \frac{24}{(Re)_p}$$

where F is the drag force/unit projected area and $(Re)_p = \dfrac{Du\rho}{\mu}$

(First liquid separates A and B from C. A and B separated by liquid with density $> 725 \text{ kg/m}^3$)

Problem 3-8: A continuous centrifugal classifier is to be used to clarify an aqueous suspension containing particles of 0.05 mm diameter, density 2500 kg/m^3. The bowl of the classifier contains an outer layer, 20 mm thick, of thickened slurry inside which is a 40 mm layer of the suspension. If the diameter of the classifier bowl is 400 mm and it rotates at 25 Hz what is the minimum time the suspension must stay in the classifier?

Assume the drag force on the particles, (diameter D) which may be assumed spherical, is given by

$$F = C_D A \rho u^2$$

where

$$C_D = \frac{9.25}{(Re)_p^{0.6}} \quad A = \pi D^2/4 \text{ and } (Re)_p = (Du\rho/\mu)$$

$$(0.05 \text{ s})$$

4. Basic Heat Transfer

4.1 Introduction

One of the problems which frequently complicate the study of heat transfer processes arises from the observation that the behaviour which such processes exhibit is invariably the result of a number of effects occurring simultaneously. The basic processes of heat transfer by conduction, convection and radiation are conveniently studied in isolation; however, in any particular engineering problem one or more of these phenomena may be important.

4.2 Heat transfer processes

4.2.1 Conduction

If temperature gradients exist within a solid body the heat will be transferred from regions of high temperature to regions of lower temperature. This phenomenon is called thermal conduction. The thermal energy may be transferred by means of electrons which are free to move through the lattice structure of the material. In addition, or alternatively, it may be transferred as vibrational energy in the lattice structure. Whilst a knowledge of these mechanisms is essential if the conduction characteristics of a particular material are to be deduced from its molecular structure, it is unnecessary when considering thermal conduction as an engineering problem.

Consider heat transfer through a plane wall, area A, thickness L, where the surfaces of the wall are maintained at temperatures T_1 and T_2 respectively, $(T_1 > T_2)$. For many engineering materials the heat flow Q through the wall is proportional to the area of wall, the temperature difference across it, and inversely proportional to the thickness of the wall, that is

$$Q \propto A(T_1 - T_2)/L$$

or

$$Q = kA(T_1 - T_2)/L \qquad (4\text{-}1)$$

where k is a property of the material of the wall called the thermal conductivity. Equation 4-1 is frequently called Fourier's Law, after the celebrated French mathematical physicist who made great contributions to the solution of the mathematical problems associated with conduction heat transfer. Substitution of the appropriate SI units into eqn. 4-1 gives the SI unit of conductivity, namely W/mK.

If we consider conduction across a small planar element of a body then eqn. 4-1 can be written in differential form

$$Q_n = -k_n A_n \frac{dT}{dn} \qquad (4\text{-}2)$$

In this equation, Q_n and dT/dn are the heat flowrate and temperature gradient in the direction n respectively, and A_n is the area exposed to the heat flow normal to the direction n. The value of thermal conductivity k_n should also be that applicable to the direction of heat flow since the material might not be isotropic (for example wood exhibits a higher conductivity measured in the direction of the grain, than that measured across the grain). The treatment of problems of conduction in anisotropic materials is extremely complex and is beyond the scope of this text.

4.2.2 Convection

If heat is transferred between a solid surface and an adjacent fluid, the heat flow at that surface is conveniently described by the equation

$$Q = hA \, \Delta T \qquad (4\text{-}3)$$

Here Q is the heat flow at the surface (W), A the area exposed to heat transfer (m^2) and ΔT the temperature difference (K) between the surface and the fluid.

The proportionality factor h is called the heat transfer coefficient

(W/m^2 K). Equation 4-3 is sometimes known as 'Newton's Law of Cooling'. However, this is misleading since it is not really a law but rather an equation defining the heat transfer coefficient h. Unlike thermal conductivity, h is not simply a property of the fluid involved in the convection process but depends also upon the geometry of the system involved, the dynamics of the flow, and the prevailing thermal conditions. This equation finds quite general application and is used to describe all forms of convective process including those involving change of ph ase for example, boiling, and condensation.

4.2.3 Radiation

The third mechanism by which heat (or thermal energy) is transferred to or from a surface is by means of electromagnetic radiation. Thermal radiation occupies only a small section of the electromagnetic spectrum in the wavelength range $0 \cdot 1-100 \, \mu$m. This includes the whole of the visible wavelengths and sections of infrared and ultraviolet spectra. The basic rate equations for radiative transfer are based on the Stefan—Boltzmann Law for the so-called black body, viz

$$E_b = \sigma_b T^4 \tag{4-4}$$

where E_b is the energy radiated/unit time, unit area (W/m^2) and T the absolute temperature (K) of the body. The constant σ_b is the Stefan—Boltzmann constant, and

$$\sigma_b = 5 \cdot 669 \times 10^{-8} \text{ W/m}^2\text{K}^4$$

The problem of radiation will not be considered further in this chapter. For a detailed discussion of the fundamentals of radiation processes the reader is referred to the further reading section.

4.3 The measurement of thermal conductivity

In view of the wide range of materials used and processed by industry it is not surprising to find that virtually all values of thermal conductivity used in design calculations have been determined experimentally. Some of the available sources of this data have been outlined in Chapter 1.

Several methods of measurement are now regarded as standard, the actual technique depending on the material under investigation.

If the material is a metal, thermal conductivity is determined by heating (usually by electrical means) one end of a cylindrical bar of the material, and cooling the other with a stream of water. Thermocouples fitted along the axis of the bar permit temperatures at adjacent points T_1 and T_2, separated by distance L, to be measured. If, under steady-state conditions, the actual heat flow along the bar Q (which is estimated from the electrical input power with allowances made for heat losses, etc.) and an average temperature gradient $(T_1 - T_2)/L$ are measured, then k can be calculated using ·eqn. 4-1. This will be an average value of k over the range T_1 to T_2, however $(T_1 - T_2)$ can be made sufficiently small that this value is appropriate to the mean temperature $(T_1 + T_2)/2$. An alternative form of this method is discussed in Example 4-1.

When the material to be tested is a non-metallic solid then the value of k is determined usually by the guarded hot plate technique. In this case, the test specimen is made in the form of a thin plate, thickness L, sandwiched between two plates, one of which is heated electrically, the other being water cooled. In order to prevent edge losses from the main heated plate, thus ensuring a uniform temperature over its surface, a special separately controlled heater is fitted along the plate edges. As before, if the net heat flow through the specimen can be determined, together with its surface temperatures, T_1 and T_2, then an average value of k can be calculated using eqn. 4-1. It is useful and instructive to consider why two methods, differing only in specimen geometry, are required for metallic and non-metallic solids.

The accurate measurement of thermal conductivity of liquids and gases is more complicated because great care must be taken to ensure that a true value of k is measured, and not one enhanced by the effects of natural convection within the test apparatus. Such effects can be minimized by the use of an apparatus which consists of a fine wire, suspended along the axis of a small bore tube. The specimen fluid is contained in the annular region between the wire and the tube. The determination of k consists of measuring the electrical input to the wire necessary to maintain a given temperature difference between the wire and the tube wall.

Many other methods have been devised for the measurement of k. Descriptions of these techniques are to be found in many of the references, which usually accompany tabulated k values.

Example 4-1: The thermal conductivity of solid materials is to be measured as shown in Fig. 4-1. A small disc, thickness t, of the material under test is sandwiched between two cylinders of the same diameter, which are made known from a material of known conductivity K. The composite cylinder fits into a housing designed to prevent heat losses from the curved surfaces of the cylinder. The housing is also equipped with a heat source and sink as shown. During a test, the temperatures

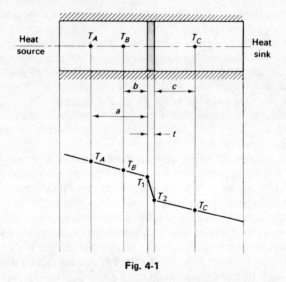

Fig. 4-1

T_A, T_B and T_C are measured by thermocouples. Sketch the temperature distribution along the cylinder, and derive an expression for the unknown conductivity k, in terms of the measured temperatures, their locations, a, b, and c, and the value K.

Solution The temperature distribution along the composite cylinder is shown in Fig. 4-1. It should be noted that this distribution presupposes that the following conditions are established:

    (a) steady-state conditions,
    (b) negligible heat losses from the curved surfaces of the cylinder, and
    (c) no contact resistance between the main cylinders and the specimen.

Under steady-state conditions, the heat flow along the cylinder, cross-sectional area A, is given by

$$Q = \frac{KA}{(a-b)}\,(T_A - T_B) \tag{4-5}$$

$$= \frac{KA}{b}\,(T_B - T_1) \tag{4-6}$$

$$= \frac{KA}{c}\,(T_2 - T_C) \tag{4-7}$$

and for the specimen

$$Q = \frac{kA}{t}(T_1 - T_2)$$ (4-8)

From eqns. 4-6 and 4-7

$$\frac{Q}{KA}(b + c) = (T_B - T_1) + (T_2 - T_C) = (T_B - T_C) - (T_1 - T_2)$$
(4-9)

From eqn. 4-5

$$\frac{Q}{KA} = \frac{(T_A - T_B)}{(a - b)}$$ (4-10)

From eqn. 4-8

$$\frac{Q}{A} = \frac{k}{t}(T_1 - T_2)$$

$$\frac{Q}{KA} = \frac{k}{Kt}(T_1 - T_2) = \frac{(T_A - T_B)}{(a - b)}$$

Therefore

$$(T_1 - T_2) = \frac{Kt(T_A - T_B)}{k(a - b)}$$ (4-11)

Substituting for Q/KA and $(T_1 - T_2)$ from eqns. 4-10 and 4-11 respectively into eqn. 4-9

$$\frac{(T_A - T_B)}{(a - b)}(b + c) = (T_B - T_C) - \frac{Kt(T_A - T_B)}{k(a - b)}$$

which gives

$$k = \frac{Kt(T_A - T_B)}{(T_B - T_C)(a - b) - (T_A - T_B)(b + c)}$$

4.4 The measurement of heat transfer coefficients

The heat transfer coefficient has been defined by eqn. 4-3 and therefore an experimental determination of this parameter involves the measurement of the rate of heat flow Q, through a surface of area A, where the difference in temperature between the surface and the adjacent fluid is ΔT. In addition to this essentially direct method of

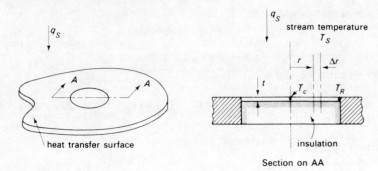

Section on AA

Fig. 4-2

measurement numerous indirect methods have been devised as a brief glance at current heat transfer literature will reveal. The measurement of heat flow rate and heat transfer coefficient has also proved to be a topic which stimulates invention, and many ingenious heat transfer meters have been devised. Example 4-2 considers the operation of such a device.

Example 4-2: A transducer is to be designed to measure heat flux q_s, and heat transfer coefficient h in industrial apparatus. As shown in Fig. 4-2 a tiny disc radius R, thickness t, is to be mounted flush with the heat transfer surface in a suitable mount. The back of the disc is assumed to be perfectly insulated so that heat flux q_s (radiative and/or convective) reaching the disc, will flow radially across the disc to the surrounding surface. Provision will be made to measure the temperature of the disc at its centre T_c, the temperature difference across the radius of the disc, $(T_c - T_R)$, and also the temperature difference between the fluid stream and the disc. By considering the heat balance on an annular element of the disc show that the temperature distribution in the disc obeys the equation

$$r\frac{d^2T}{dr^2} + \frac{dT}{dr} = -\frac{q_s r}{kt}$$

Show that

$$h = \frac{4kt(T_c - T_R)}{\Delta T_s R^2}$$

where $\Delta T_s = T_s - T_c$, and T_s is the fluid stream temperature.

Solution Consider the heat balance on an annular element of the disc between radii r and $r + \Delta r$:

90

Heat flow/unit time into the element by conduction
= heat flux x area = $q_r 2\pi rt$
Heat flow/unit time into the element by convection = $q_s 2\pi r \Delta r$
Heat flow/unit time out of the element by conduction
= $q_{r+\Delta r} 2\pi (r + \Delta r)t$.
Under steady-state conditions a simple heat balance gives

$$\underbrace{q_r 2\pi rt + q_s 2\pi r \Delta r}_{\text{(Heat in)}} = \underbrace{q_{r+\Delta r} 2\pi (r + \Delta r)t}_{\text{(Heat out)}}$$

so

$$\frac{q_{r+\Delta r}(r + \Delta r) - q_r r}{\Delta r} = \frac{q_s r}{t}$$

Taking the limit as $\Delta r \to 0$

$$\frac{q_{r+\Delta r}(r + \Delta r) - q_r r}{\Delta r} \to \frac{d}{dr}(qr)$$

thus

$$\frac{d}{dr}(q \cdot r) = \frac{q_s r}{t}$$

Using the differential form of Fourier's Law $q = -k(dT/dr)$

$$\frac{d}{dr}\left(-k \cdot \frac{dT}{dr} \cdot r\right) = \frac{q_s r}{t}$$

if k is a constant, then

$$\frac{d}{dr}\left(\frac{dT}{dr} \cdot r\right) = r\frac{d^2 T}{dr^2} + \frac{dT}{dr} = -\frac{q_s r}{kt}$$

integration gives

$$r\frac{dT}{dr} = -\frac{q_s r^2}{2 kt} + C_1$$

therefore

$$\frac{dT}{dr} = -\frac{q_s r}{2 kt} + \frac{C_1}{r}$$

Since dT/dr must always be finite, or zero for $r = 0$ then C_1 must be zero.

In general, the symmetry of the system indicates that $dT/dr = 0$ for

$r = 0$. A further integration gives

$$T = -\frac{q_s r^2}{4kt} + C_2$$

When $r = R$, $T = T_R$ then

$$C_2 = T_R + \frac{q_s R^2}{4\ kt}$$

and when $r = 0$, $T = T_c$ then

$$T_c = C_2 = T_R + \frac{q_s R^2}{4\ kt}$$

Hence

$$q_s = \frac{4kt(T_c - T_R)}{R^2}$$

Assuming $T_c - T_R \ll T_s - T_c$, then

$$h = \frac{q_s}{(T_s - T_c)} = \frac{4\ kt(T_c - T_R)}{\Delta T_s R^2}$$

Example 4-3: The inner surface of a high temperature reactor will operate at 1623 K. The wall of the reactor will have an overall thickness of 350 mm and is to be made up of an inner layer of firebrick material ($k_r = 0.86$ W/m K), covered with a layer of insulation ($k_i = 0.16$ W/m K). This insulating material has a maximum operating temperature of 1473 K. The ambient temperature will be 293 K and it is estimated that the heat transfer coefficient at the exposed surface of the insulation will be 10 W/m²K. Calculate the thickness of refractory and insulation which gives minimum heat loss, and the magnitude of this loss in W/m². Also calculate the surface temperature of the insulation. If the calculated heat loss is unacceptable, would the addition of another layer of insulation be a satisfactory solution? Give your reasons.

Solution A section through the reactor wall is shown in Fig. 4-3. In order to minimize heat loss through the wall, the layer of insulating material should be as thick as possible. However, this thickness is limited by the condition that the material shall not be exposed to temperatures in excess of 1473 K. Therefore the refractory/insulation inferface temperature will be set at 1473 K.

Let the heat flux through the wall be q W/m². Under steady state conditions this will be constant throughout the wall, so for the refractory

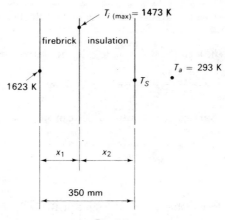

Fig. 4-3

$$q = \frac{k_r(T_1 - T_i)}{x_1}$$

for the insulation

$$q = \frac{k_i(T_i - T_s)}{x_2} \qquad (4\text{-}12)$$

and convection from insulation surface

$$q = h(T_s - T_a) \qquad (4\text{-}13)$$

Therefore

$$q\left(\frac{x_1}{k_r} + \frac{x_2}{k_i} + \frac{1}{h}\right) = (T_1 - T_a)$$

hence

$$q = \frac{(T_1 - T_a)}{\dfrac{x_1}{k_r} + \dfrac{x_2}{k_i} + \dfrac{1}{h}} = \frac{k_r}{x_1}(T_1 - T_i) \qquad (4\text{-}14)$$

Since $x_1 = (350 - x_2)$ mm then

$$\frac{(1623 - 293)}{\dfrac{(350 - x_2)}{1000 \times 0\cdot86} + \dfrac{x_2}{1000 \times 0\cdot16} + \dfrac{1}{10}} = \frac{0\cdot86\,(1623 - 1473)\,100}{(350 - x_2)}$$

which gives $1986\,x_2 = 40\,010$

Thus, thickness of insulation $x_2 = 201 \cdot 5$ mm, and thickness of refractory $(350 - x_2) = 148 \cdot 5$ mm.

The heat flow through the wall can be calculated from eqn. 4-14

$$q = \frac{0 \cdot 86 \, (1623 - 1473) \, 1000}{148 \cdot 5} = 869 \text{ W/m}^2$$

The insulation surface temperature T_sK, can now be calculated from eqn. 4-13

$$q = 10(T_s - 293) = 869$$

$$T_s = 86 \cdot 9 + 293 = 379 \cdot 9 \text{ K}$$

Check on interface temperature T_i: from eqn. 4-12

$$q = \frac{0 \cdot 16 \, (T_i - 379 \cdot 9) \, 1000}{201 \cdot 5} = 869$$

and

$$T_i = 1473 \text{ K}$$

The addition of a further layer of insulating material would reduce the heat loss from the wall, however it would also reduce the temperature drop across the refractory lining and therefore T_i would increase. Since this was set initially to the maximum permissible value this solution would not be satisfactory.

Example 4-4: Laminated sheets are constructed from single sheets of a plastic using a suitable adhesive. When a number of sheets has been assembled they are clamped between steel plates to prevent distortion as the adhesive hardens. The steel plates are maintained at a temperature of 343 K to dissipate heat evolved during the hardening process. At maximum, this heat evolution is equivalent to a steady uniform heat generation of 100 W/m³. How many laminated sheets, which are 3 mm thick, can be processed at any one time if the maximum temperature in the stack must not exceed 353 K? The effective thermal conductivity of the stack may be taken as 0.2 W/m K. Heat transfer in the stack may be assumed to be one dimensional.

Solution The general conduction equation written for the rectangular co-ordinate system is appropriate to this problem, namely

$$\frac{\partial^2 T}{\partial x^2} + \frac{\partial^2 T}{\partial y^2} + \frac{\partial^2 T}{\partial z^2} + \frac{H}{k} = \frac{1}{\alpha} \frac{\partial T}{\partial \tau}$$

Since steady-state conditions are assumed to prevail then

$$T \neq f(\tau)$$

therefore

$$\frac{\partial T}{\partial \tau} = 0$$

Furthermore, where significant variations in temperature are assumed to exist in only one direction, in this case through the stack, say the x direction, then

$$T = f(x) \text{ only}$$

so

$$\frac{\partial T}{\partial y} = 0$$

and

$$\frac{\partial T}{\partial z} = 0 \text{ for all } y \text{ and } z$$

hence

$$\frac{\partial^2 T}{\partial y^2} = \frac{\partial^2 T}{\partial z^2} = 0$$

and the general equation reduces to

$$\frac{d^2 T}{dx^2} + \frac{H}{k} = 0$$

Integration of this equation is straightforward since H and k are assumed constant

$$\frac{dT}{dx} = -\frac{Hx}{k} + C_1$$

and

$$T = -\frac{Hx^2}{2k} + C_1 x + C_2$$

The integration constants C_1 and C_2 must now be determined from the boundary conditions for this particular problem. The stack of laminated sheets is shown in Fig. 4-4. The stack is $2L$ m thick and the origin

Fig. 4-4

of the co-ordinate system will be taken as the centre of the stack. Since both surfaces of the stack are at the same temperature then the temperature profile through the stack will be symmetrical* about $x = 0$

$$\frac{dT}{dx} = 0 \quad \text{for } x = 0$$

therefore

$$C_1 = 0$$

It follows that when $x = 0$

$$T = T_{max} = 353 \text{ K}$$

and

$$C_2 = 353$$

When $x = L$

$$T = 343$$

so

$$343 = \frac{-HL^2}{2k} + 353$$

---

*As an alternative approach the following boundary conditions may be used

$$T = 343 \text{ K at } x = L$$

$$T = 343 \text{ K at } x = -L$$

Show that $dT/dx = 0$ for $x = 0$ and then proceed as before. (This approach will be necessary in Problem 4-12.)

and

$$L^2 = \frac{2k \times 10}{H} = \frac{2 \times 0.2 \times 10}{100} = 0.04$$

therefore

$L = 0.2$ m

Maximum stack depth is

$2L = 0.4$ m

Approximate number of laminated sheets in the stack is

$$\frac{0.4 \times 1000}{3} = 133.3$$

This result indicates that the maximum number of sheets should be limited to 133. In practice, it would be essential to ensure that sufficient pressure is applied to the stack in order to minimize thermal contact resistance between the laminated sheets.

Example 4-5: A flat plate fuel element for a nuclear reactor is 7 mm thick and is clad on each face with aluminium 2 mm thick. The rate of heat generation is uniform within the element and has a magnitude of 3×10^4 W/kg of uranium. Determine the temperatures at the free surface of the aluminium, the aluminium/uranium interface, and at the centre of the fuel element. The coolant temperature is 413 K.
   Additional data:

Density of uranium 18.9×10^3 kg/m^3
Thermal conductivity of uranium 24.3 W/m K
Thermal conductivity of aluminium 2.1×10^2 W/m K
The heat transfer coefficient at the aluminium/coolant interface 2.84×10^4 W/m^2 K.

Solution   A sketch of the arrangement of the fuel element is shown in Fig. 4-5. Let the total heat generated within the uranium per unit plate area be $2Q$

$$2Q = \underbrace{(1 \times 7 \times 10^{-3})(18.9 \times 10^3)}_{\text{mass of uranium/m}^2 \text{ plate area}} 3 \times 10^4 \text{ W/m}^2 = 39.7 \times 10^5 \text{ W/m}^2$$

Then, Q heat flow in each normal direction is 19.85×10^5 W/m^2. Let

97

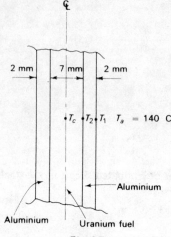

Fig. 4-5

the aluminium/coolant interface temperature be T_1 then

$$Q = h \cdot A \cdot \Delta T = h(T_1 - 413)$$

now A is $1\,m^2$, so

$$T_1 - 413 = \frac{19 \cdot 86 \times 10^5}{2 \cdot 84 \times 10^4} = 69 \cdot 9$$

then

$$T_1 = 482 \cdot 9 \text{ K}$$

Let the aluminium/uranium interface temperature be T_2

$$Q = \frac{kA\Delta T}{x} \qquad A = 1\,m^2 \qquad \Delta T = (T_2 - T_1)$$

then

$$Q = \frac{2 \cdot 1 \times 10^2 (T_2 - 482 \cdot 9)}{0 \cdot 2 \times 10^{-2}} = 19 \cdot 85 \times 10^5 \text{ W/m}^2$$

so

$$T_2 = 482 \cdot 9 + \frac{19 \cdot 85 \times 10^5 \times 0 \cdot 2 \times 10^{-2}}{2 \cdot 1 \times 10^2} = 501 \cdot 8 \text{ K}$$

Since the uranium fuel is subject to uniform internal heat generation, the temperature distribution within this material will be described by the equation used in Example 4-4, viz

$$\frac{d^2 T}{dx^2} + \frac{H}{k} = 0$$

The use of this equation assumes, as before, that the heat flow in the present system is essentially one dimensional. Now, H is the internal heat generation/unit volume

$$H = \frac{39 \cdot 7 \times 10^5}{7 \times 10^{-3}} = 56 \cdot 7 \times 10^7 \ \text{W/m}^3$$

Integration of

$$\frac{d^2 T}{dx^2} = -\frac{H}{k}$$

gives

$$\frac{dT}{dx} = -\frac{Hx}{k} + C_1$$

and

$$T = -\frac{Hx^2}{2k} + C_1 x + C_2$$

as before, the constants C_1 and C_2 are to be determined from the boundary conditions, i.e.

(i) $x = 3 \cdot 5$ mm $T = 501 \cdot 8$ K

(ii) $x = 0$ $\dfrac{dT}{dx} = 0$ (implied by the symmetry of the system)

From (ii)

$$C_1 = 0$$

From (i)

$$501 \cdot 8 = \frac{-56 \cdot 7 \times 10^7 \ (3 \cdot 5 \times 10^{-3})^2}{2 \times 24 \cdot 3} + C_2$$

from which

$$C_2 = 501 \cdot 8 + 143 = 644 \cdot 8 \ \text{K}$$

But at $x = 0$, $T = T_c$ so

$$T_c = C_2 = 644 \cdot 8 \text{ K}$$

Example 4-6: A rod is heated at one end and the heat is dissipated by conduction along the rod and convection from its surface. Show that the temperature difference θ, between the rod and the surrounding fluid, assumed to be at a constant temperature, T_g, is described by the equation

$$\frac{d^2 \theta}{dx^2} - m^2 \theta = 0$$

where $m^2 = \dfrac{hP}{kA}$

where P and A are the constant perimeter and cross sectional area of the rod respectively, k the thermal conductivity of the rod material, and h the heat transfer coefficient at the surface of the rod. For a long rod show that

$$\theta = \theta_0 \, e^{-mx} \quad \text{where } \theta = \theta_0 \text{ at } x = 0$$

A long rod exposed to air at 293 K, is heated at one end and when steady conditions are attained the temperature at two points along the rod separated by 100 mm are found to be 393 K and 373 K respectively. If the rod diameter is 20 mm and $k = 120$ W/m K, estimate the heat transfer coefficient at the surface of the rod.

Solution Consider an element of the rod as shown in Fig. 4-6. Heat flow into the element by conduction at plane XX is

$$Q_x = q_x A$$

Heat flow out of the element by conduction at plane YY is

$$Q_{x + \Delta x} = q_{x + \Delta x} A$$

Heat flow from the surface of the element by convection is

$$Q_c = h \cdot P \Delta x (T - T_g)$$

A heat balance on the element gives

$$q_x A = h P \Delta x (T - T_g) + q_{x + \Delta x} A$$

so

$$\frac{q_{x + \Delta x} - q_x}{\Delta x} = -\frac{hP}{A} \theta$$

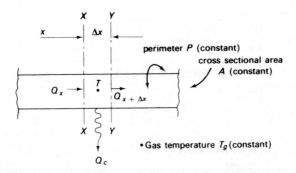

The rod temperature T is assumed to be a function of x only.

Fig. 4-6

where $\theta = (T - T_g)$, $\left(\text{note } \dfrac{d\theta}{dx} = \dfrac{dT}{dx} \right)$

Taking the limit as $\Delta x \to 0$

$$\frac{q_{x+\Delta x} - q_x}{\Delta x} \to \frac{dq}{dx}$$

therefore

$$\frac{dq}{dx} = -\frac{hP}{A}\theta$$

but

$$q = -k\frac{dT}{dx} = -k\frac{d\theta}{dx} \quad \text{(Differential form of Fourier's Law)}$$

Therefore

$$\frac{d}{dx}\left[-k\frac{d\theta}{dx} \right] = -\frac{hP}{A}\theta$$

so

$$-\frac{d^2\theta}{dx^2} = -\frac{hP}{kA}\theta$$

or

$$\frac{d^2\theta}{dx^2} - m^2\theta = 0$$

101

where

$$m^2 = \frac{hP}{kA}$$

The above differential equation presents no difficulty in solution. The general solution may be written as either

$$\theta = C_1 e^{mx} + C_2 e^{-mx}$$

or

$$\theta = C_3 \cosh mx + C_4 \sinh mx$$

In the present case, the first solution is the more convenient. For a long rod $\theta \to 0$ as $x \to \alpha$ and therefore the constant C_1 must be zero. For $x = 0$ $\theta = \theta_0$ and $\theta_0 = C_2$, hence

$$\theta = \theta_0 e^{-mx}$$

Given

$$\theta_1 = (393 - 293) = 100 \text{ when } x = x_1$$

and

$$\theta_2 = (373 - 293) = 80 \text{ when } x = x_2$$

then

$$100 = \theta_0 e^{-mx_1} \text{ and } 80 = \theta_0 e^{-mx_2}$$

or

$$\frac{100}{80} = e^{m(x_2 - x_1)}$$

but

$$(x_2 - x_1) = 100 \text{ mm} = 0.1 \text{ m}$$

so

$$1.25 = e^{0.1m}$$

from which

$$m = 2.23$$

But $\quad m^2 = \dfrac{hP}{kA}$

then

$$h = 4 \cdot 97\, k\, \frac{A}{P} = 4 \cdot 96 \times 120 \times \frac{\pi d^2 \times 100}{4(1000)^2\, \pi d} = \frac{4 \cdot 97 \times 120 \times d}{4 \times 1000}$$

Since $d = 20$ mm

$$h = 2 \cdot 98\ \text{W/m}^2\ \text{K}$$

Example 4-7: The temperature of a gas stream is to be measured by using two thermocouples attached to a tube of perimeter 40 mm and wall cross sectional area $10\ \text{mm}^2$. The tube is 200 mm long and is mounted normal to the duct wall. If the thermocouples are attached to the tube at 100 mm and 200 mm from the duct wall and indicate tube wall temperatures of 619 K and 662 K respectively, estimate the gas temperature and the duct wall temperature. The heat transfer coefficient between the tube and the gas stream is $1\ \text{W/m}^2\text{K}$ and the thermal conductivity of the tube material is 20 W/m K. Heat transfer into the exposed end of the tube can be neglected.

Solution The equation which described the temperature distribution along a rod, exposed to heat transfer at its surface, was derived in Example 4-6,

$$\frac{\text{d}^2 \theta}{\text{d}x^2} - m^2 \theta = 0$$

where

$$m^2 = \frac{hP}{k\text{A}}$$

and θ is the difference between the constant gas temperature T_g and the local tube temperature T

$$\theta = (T_g - T)$$

In this problem the solution involving hyperbolic functions will be used

$$\theta = C_1 \cosh mx + C_2 \sinh mx$$

It is useful to note that

$$\cosh mx = \frac{e^{mx} + e^{-mx}}{2} \qquad \cosh 0 = 1 \qquad \text{and}\ \frac{\text{d}}{\text{d}x}(\cosh mx) = m \sinh mx$$

$$\sinh mx = \frac{e^{mx} - e^{-mx}}{2} \qquad \sinh 0 = 0 \qquad \text{and}\ \frac{\text{d}}{\text{d}x}(\sinh mx) = m \cosh mx$$

103

In order to determine the values of the constants C_1 and C_2 it is necessary to specify appropriate boundary conditions. Put

$$\theta = \theta_0 = (T_g - T_0)$$

At $x = 0$

$$\theta_0 = C_1$$

Since the heat flow Q_L into the exposed end of the tube may be neglected, then

$$Q_L = - kA \left. \frac{dT}{dx} \right|_{x=L} = 0$$

Since both k and A are finite then

$$\frac{dT}{dx}_{x=L} = 0 = \left. \frac{d\theta}{dx} \right|_{x=L} \quad (\text{since } \theta = (T_g - T))$$

and

$$\left(\frac{d\theta}{dx} \right)_{x=L} = m\theta_0 \sinh mL + C_2\, m \cosh mL = 0$$

then

$$C_2 = - \theta_0 \tanh mL$$

and

$$\theta = \theta_0 (\cosh mx - \tanh mL \sinh mx)$$

Now, for

$$x = L = 200 \text{ mm}, \quad T = 662 \text{ K and } \theta_L = (T_g - 662), \text{ so}$$

$$(T_g - 662) = \theta_0 (\cosh mL - \tanh mL \sinh mL)$$

therefore

$$(T_g - 662) = \frac{\theta_0}{\cosh mL} (\cosh^2 mL - \sinh^2 mL) = \frac{\theta_0}{\cosh mL}$$

Now

$$m = \left(\frac{hP}{kA} \right)^{0.5} = \left[\frac{1 \times 40 \times 10^{-3}}{20 \times 10 \times 10^{-6}} \right]^{0.5} = (200)^{\frac{1}{2}} = 14 \cdot 14$$

so

$$mL = 14 \cdot 15 \times 200 \times 10^{-3} = 2 \cdot 83$$

and

$$\cosh mL = \cosh 2\cdot 83 = 8\cdot 47$$

Hence

$$T_g - 662 = \frac{\theta_0}{8\cdot 47} = 0\cdot 118\, \theta_0$$

Similarly for $x = 100$ mm, $\quad T = 619$ K and

$$T_g - 619 = \theta_0 \,(\cosh 1\cdot 41 - \tanh 2\cdot 83 \sinh 1\cdot 41)$$
$$= \theta_0 \,(2\cdot 17 - 0\cdot 99 \times 1\cdot 92) = 0\cdot 27\, \theta_0$$

Hence

$$\frac{T_g - 662}{T_g - 619} = \frac{0\cdot 118}{0\cdot 27} = 0\cdot 47$$

therefore

$$T_g = 695\cdot 8\text{ K}$$

also

$$\theta_0 = \frac{695\cdot 8 - 662}{0\cdot 118} = 286$$

but

$$\theta_0 = T_g - T_0 \text{ where } T_0 \text{ is the duct wall temperature}$$

so

$$T_0 = 695\cdot 8 - 286 = 409\cdot 8\text{ K}$$

Example 4-8: The product from a chemical process is in the form of pellets which are approximately spherical with a mean diameter d of 4 mm. These pellets are initially at 403 K and must be cooled before entering a storage vessel. It is proposed to cool these pellets to the required temperature (343 K max) by passing them down a slightly inclined channel which has a porous base allowing the pellets to be fluidized into a shallow layer by a stream of air. The air will have a mean temperature of 323 K whilst passing through the bed. If the length of the channel is limited to 3 m calculate the maximum velocity of pellets along the channel. Heat transfer from the pellet surface to the air stream may be considered to be the limiting process with $hd/k_a = 2$ where h is the heat transfer coefficient at the pellet surface.

Other data: Pellet material — density = 480 kg/m³ specific heat = 2 kJ/kg K.
Thermal conductivity of air, $k_a = 0\cdot13$ W/m K.

Solution The required maximum velocity will be governed by the time required for pellet cooling, t_c. Consider a single pellet. If the internal thermal resistance of a pellet is small compared with the thermal resistance to heat transfer at its surface, the temperature gradients within the pellet can be ignored and we may write

$$\text{rate of loss of internal energy of the pellet} = -mc_p \frac{dT}{dt}$$

This loss is due to convection from the pellet surface given by $h\,A(T - T_a)$. In these equations, T is the pellet temperature and T_a is the air stream temperature; the latter is assumed to be constant, so

$$h = 2k_a/d = 2 \times 0\cdot13/(4 \times 10^{-3}) = 65 \text{ W/m}^2 \text{ K}$$

now

$$-mc_p \frac{dT}{dt} = hA(T - T_c)$$

putting $\theta = T - T_a$ gives

$$\frac{d\theta}{dt} = -\frac{hA}{mc_p}$$

or

$$d\theta = -\left(\frac{hA}{mc_p}\right) dt$$

Integration gives

$$\ln \frac{\theta_2}{\theta_1} = -\frac{hA}{mc_p} (t_2 - t_1)$$

when $t = t_1 = 0$, $\theta = \theta_1 = 403 - 323 = 80$

$t = t_2 = t_c$, cooling time $\theta = \theta_2 = (343 - 323) = 20$

and

$$\ln \frac{20}{80} = -\left[\frac{65 \times \pi d^2 \times 10^{-6}}{480 \times \dfrac{\pi d^3}{6} \times 10^{-9} \times 2 \times 10^3} \right] t_c$$

where t_c is now in seconds.

Now

$$\ln 0\cdot 25 = \ln 2\cdot 5 + \ln 10^{-1} = 0\cdot 9163 - 2\cdot 3026 = -1\cdot 3863$$

so, putting $d = 4$ mm

$$t_c = \frac{1\cdot 3863 \times 480 \times 4 \times 2}{65 \times 6} = 13\cdot 6 \text{ s.}$$

Maximum pellet velocity

$$V_m = \text{channel length/cooling time} = 3/13\cdot 6 = 0\cdot 22 \text{ m/s}$$

Example 4-9:   Water from a storage tank is supplied to a process by means of a pipeline 650 m long, 100 mm outside diameter. Under the most severe weather conditions ambient temperature and storage tank temperature are expected to have minimum values of 263 K and 278 K respectively. Estimate the minimum flowrate (kg/s) between the storage tank and the process necessary to prevent ice formation in the pipeline. The pipeline is insulated with a layer of magnesia ($k = 0\cdot 1$ W/m K) 25 mm thick, and a heat transfer coefficient of $1\cdot 2$ W/m²K at the surface of the insulation is anticipated. The specific heat of water is $4\cdot 18$ kJ/kg K. Density of water $1\cdot 0 \times 10^3$ kg/m³.

Solution   As a first approximation, it is appropriate to consider the conditions which prevail when the water temperature at the process end of the pipeline is just 273 K.

Sensible heat lost be the process water is $mc_p \Delta T_w$

where m = mass flowrate of water (kg/s)

c_p = specific heat J/kg K

ΔT_w = temperature drop $(278 - 273)$ K

This heat loss is due to heat leakage through the insulation which may be written as

$$Q_L = U_0 A_0 \theta_m$$

where A_0 is the external surface area of the insulation.

U_0 equivalent overall heat transfer coefficient based on the area A_0.

θ_m the mean temperature difference between the process water and the ambient conditions.

A cross sectional view of the pipeline and insulation is shown in

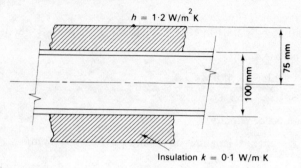

$$h = 1 \cdot 2 \text{ W/m}^2 \text{K}$$

Insulation $k = 0 \cdot 1$ W/m K

Fig. 4-7

Fig. 4-7. If the heat transferred through this composite wall is Q/unit length, then under steady-state conditions

$$Q = 2\pi r_1 h_1 (T_w - T_1) = \frac{2\pi k_w}{\ln(r_2/r_1)} (T_1 - T_2) = \frac{2\pi k_w}{\ln(r_3/r_2)} (T_2 - T_3)$$

$$= 2\pi r_3 h_0 (T_3 - T_a)$$

Then

$$\frac{Q}{2\pi} \left[\frac{1}{r_1 h_1} + \frac{\ln(r_2/r_1)}{k_w} + \frac{\ln(r_3/r_2)}{k_i} + \frac{1}{r_3 h_0} \right] = (T_w - T_a)$$

from which

$$Q = \frac{2\pi(T_w - T_a)}{\left[\frac{1}{r_1 h_1} + \frac{\ln(r_2/r_1)}{k_w} + \frac{\ln(r_3/r_2)}{k_i} + \frac{1}{r_3 h_0} \right]}$$

Putting

$$Q = 2\pi r_3 \, U_0 (T_3 - T_a)$$

$$\frac{1}{U_0} = \left[\frac{r_3}{r_1 h_1} + \frac{r_3 \ln(r_2/r_1)}{k_w} + \frac{r_3 \ln(r_3/r_2)}{k_i} + \frac{1}{h_0} \right]$$

The first two terms on the right hand side of this equation represent the thermal resistances offered by convective heat transfer from the process stream to the inside surface of the pipeline, and conduction through the pipe wall itself, respectively. In many instances these two terms will be negligible, when compared with the thermal resistance offered by the insulation, and at the exposed surface of the insulation.

This will be taken as an appropriate assumption here, therefore

$$\frac{1}{U_0} = \frac{75 \times 10^{-3} \ln(75/50)}{0 \cdot 1} + \frac{1}{1 \cdot 2} = 1 \cdot 137$$

or

$$U_0 = 0 \cdot 88 \text{ W/m}^2 \text{ K}$$

In this case, the appropriate value for the mean temperature difference θ_m is the logarithmic mean temperature difference ($LMTD$), and

$$\theta_m = LMTD = \frac{\theta_1 - \theta_2}{\ln(\theta_1/\theta_2)}$$

so

$$\theta_1 = 278 - 263 = 15, \quad \theta_2 = 273 - 263 = 10$$

and

$$\theta_m = \frac{15 - 10}{\ln 1 \cdot 5} = 12 \cdot 3$$

External surface area of the insulation is

$$\pi \times 15 \times 10^{-2} \times 650 = 306 \cdot 3 \text{ m}^2$$

Heat loss through the insulation is

$$U_0 A_0 \theta_m = 0 \cdot 88 \times 306 \cdot 3 \times 12 \cdot 3 = 3315 \cdot 4 \text{ W} = m c_p \Delta T_w$$

Therefore

$$m \times 4 \cdot 18 \times 10^3 \times 5 = 3315 \cdot 4$$

from which

$$m = 0 \cdot 158 \text{ kg/s}$$

Example 4-10: Show that the addition of fins of uniform cross sectional area A, perimeter P, and length L, will increase the rate of heat transfer from a surface, provided

$$(kP/hA)^{\frac{1}{2}} \tanh mL > 1$$

where $m = (hP/kA)^{\frac{1}{2}}$ k is the thermal conductivity of the fin material, and h is the heat transfer coefficient for the fins and the bare surface. Tip losses from the fins may be neglected.

109

For a steel fin ($k = 42$ W/m K), 5 mm diameter and 25 mm long compare the performance for heat transfer coefficients of 5 and 10 000 W/m^2 K respectively. Comment on the results.

Solution The temperature distribution for a fin of constant cross sectional area has already been derived in Example 4-6 and the solution will be written in terms of hyperbolic functions, that is

$$\theta = A \cosh mx + B \sinh mx$$

As before,

$$\theta = (T - T_a)$$

where $T(= f(x))$ is the fin temperature and T_a is the constant ambient temperature. The constants A and B are determined from the boundary conditions.

at $\quad x = 0, \quad T = T_0, \quad \theta = \theta_0 = T_0 - T_a$ therefore $A = \theta_0$.

At $\quad x = L, \quad \left. \dfrac{dT}{dx} \right|_{x = L} = 0$

but

$$\frac{dT}{dx} = \frac{d\theta}{dx}$$

therefore

$$\frac{d\theta}{dx_{x = L}} = m\theta_0 \sinh mL + mB \cosh mL = 0$$

from which

$$B = -\theta_0 \tanh mL \text{ and } \theta = \theta_0 (\cosh mx - \tanh mL \sinh mx)$$

The heat loss Q_f from the surface of the fin can be calculated in two ways:

(a) Heat loss from an element of the fin surface

$$dQ_f = h \times \text{area of the element} \times \text{temperature difference} = h \cdot P dx \cdot \theta.$$

Integrating over the fin length we have

$$Q_f = \int_0^L hP\theta_0 (\cosh mx - \tanh mL \sinh mx) \, dx$$

and (b) Heat conducted into the fin at its root ($x = 0$)

$$Q_f = -kA \left. \frac{dT}{dx} \right|_{x=0}$$

then

$$\left. \frac{dT}{dx} \right|_{x=0} = \left. \frac{d\theta}{dx} \right|_{x=0} = -m\theta_0 \tanh mL$$

and

$$Q_f = kAm\theta_0 \tanh mL$$

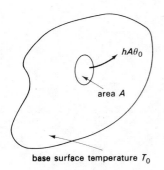
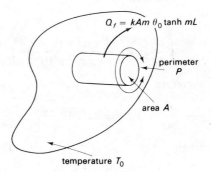

Fig. 4-8

Figure 4-8 shows the base heat transfer surface before and after the fin has been attached. The heat transfer through the root area A before fin attachment

$$Q = hA\theta_0$$

after fin attachment the heat transfer through the root area A is Q_f

$$Q_f = kAm\theta_0 \tanh mL$$

If the fin is to improve heat dissipation from the primary surface then

$$\frac{Q_f}{Q} \text{ must be} > 1$$

or

$$\frac{Q_f}{Q} = \frac{kAm\theta_0 \tanh mL}{hA\theta_0} = (kP/hA)^{1/2} \tanh mL \text{ must be} > 1$$

111

Consider the steel fin when $h = 5 \text{ W/m}^2 \text{ K}$

$$m = \left(\frac{hP}{kA}\right)^{1/2} = \left(\frac{5\pi d \times 10^{-3} \times 4}{42\pi d^2 \times 10^{-6}}\right)^{1/2} = 9\cdot75 \quad \text{for } d = 5 \text{ mm}$$

Therefore

$$mL = \frac{9\cdot75 \times 2\cdot5}{100}$$

and

$$\tanh mL = 0\cdot238$$

Thus

$$\left(\frac{kP}{hA}\right)^{1/2} = \left(\frac{42 \times \pi d \times 10^{-3} \times 4}{5 \times \pi d^2 \times 10^{-6}}\right)^{1/2} = 82$$

therefore

$$\frac{Q_f}{Q} = 0\cdot238 \times 82 = 19\cdot52$$

When

$$h = 10\,000 \text{ W/m}^2 \text{ K}$$

$$m = \left(\frac{hP}{kA}\right)^{1/2} = \left(\frac{10\,000 \times \pi d \times 10^{-3} \times 4}{42 \times \pi d^2 \times 10^{-6}}\right)^{1/2} = 4\cdot36 \times 10^2$$

and

$$mL = 4\cdot35 \times 10^2 \times \frac{2\cdot5}{100} = 10\cdot87$$

and

$$\tanh mL \doteqdot 1$$

and

$$\left(\frac{kP}{hA}\right)^{1/2} = \left(\frac{42 \times \pi d \times 10^{-3} \times 4}{10\,000 \times \pi d^2 \times 10^{-6}}\right)^{1/2} = 1\cdot83$$

therefore

$$\frac{Q_f}{Q} = 1\cdot83$$

List of symbols

A	area	(m^2)
c_p	specific heat	$(J/kg\ K)$
d	diameter	(m)
E_b	radiant energy of a black body	(W/m^2)
h	heat transfer coefficient	$(W/m^2\,K)$
k, K	thermal conductivity	$(W/m\ K)$
L	linear dimension	(m)
m	parameter $(hP/KA)^{0.5}$	$(1/m)$
P	perimeter	(m)
Q	heat flow	(W)
q	heat flux	(W/m^2)
r, R	radius	(m)
T	temperature	(K)
U	overall heat transfer coefficient	$(W/m^2\ K)$
T	velocity	(m/s)
x, y, z	rectangular co-ordinates	(m)
α	thermal diffusivity	(m^2/s)
θ	temperature difference	(K)
θ_m	logarithmic mean temperature difference	(K)
σ_b	Stefan–Boltzmann constant	$(W/m^2\,K^4)$
τ	time	(s)

Further reading

GEBHART, B. *"Heat Transfer"* 2nd Edition McGraw-Hill, New York (1971).

BENNETT, C. O. & MYER, J. E. *"Momentum, Heat and Mass Transfer"* McGraw-Hill, New York (1962).

CHAPMAN, A. J. *"Heat Transfer"* 2nd Edition Macmillan, New York (1967).

KREITH, F. *"Principles of Heat Transfer"* Feffer and Simons, Scranton, Pennsylvania, (1965).

JAKOB, M. *"Heat Transfer"* Vol. 1 Wiley, New York (1949).

GROBER, H., ERK, E. and GRIGULL, U. *"Fundamentals of Heat Transfer"* McGraw-Hill, New York (1961).

BIRD, R. B., STEWART, W. E. & LIGHTFOOT, E. N. *"Transport Phenomena"* Wiley, New York (1960).

SCHNEIDER, P. J. *"Conduction Heat Transfer"* Addison-Wesley, Cambridge, Mass. (1955).

KRAUS, A. D. *"Extended Surfaces"* Cleaver-Hume, London (1964).
KAY, W. M. *"Convective Heat and Mass Transfer"* McGraw Hill, New
York (1966).

Problem 4-1: The plane wall of a furnace is constructed from two
layers of different materials. The inner layer is made from firebrick,
150 mm thick, whilst the outer layer is ordinary brick, again 150 mm
thick. The thermal conductivities of firebrick and ordinary brick are
1·7 W/m K and 0·8 W/m K respectively. Under steady conditions, the
inside wall temperature is 1023 K, and the outer brick surface is at
383 K. To reduce heat losses an 80 mm layer of magnesia (k = 0·09)
W/m K) is added to the external surface of the wall. Under this
insulated condition, the inside wall temperature becomes steady at
1053 K whilst the outer surface settles to 338 K. Calculate:

    (a) the heat loss through the wall, before and after the addition
        of the magnesia insulation, and
    (b) the following interface temperatures for the insulated wall
        (i) firebrick − ordinary brick
        (ii) ordinary brick − magnesia.

$$(2321 \text{ W/m}^2, 614 \text{ W/m}^2 \text{ } 999 \text{ K}, 884 \text{ K})$$

Problem 4-2: A steam pipe 140 mm outside diameter is to be covered
with two layers of insulation each 40 mm thick, the thermal conductivity of one material being four times that of the other. Show that the
effective conductivity of the layers is approximately 20% less when the
better insulator is on the inside.

Problem 4-3: A pipe 20 mm bore, 30 mm outside diameter carries a
fluid at 448 K when the ambient temperature is 293 K. The inner and
outer heat transfer coefficients are 22 and 5 W/m²K. Calculate the heat
loss from the pipe in W/m. What will the heat loss be if a layer of
insulation, 50 mm thick ($k = 0·8$ W/m K) is applied to the pipe. Ignore
the thermal resistance at the pipe wall. Comment on the results.

$$(54·5 \text{ W/m}, 102·9 \text{ W/m})$$

Problem 4-4: A wire 3 mm diameter carries a maximum current of
12 amperes. The wire has an electrical resistance of 0·026 ohm/m and
is insulated with a material which has a thermal conductivity of
0·08 W/m K. What thickness of insulation will give rise to minimum
wire temperature? Temperature gradients within wire may be ignored.
The heat transfer coefficient at the insulation surface is 20 W/m²K.

Plot the variation of wire temperature with insulation thickness from 0·5 to 3·5 mm. Discuss this characteristic.

(4 mm)

Problem 4-5: Describe briefly an indirect method of measuring heat transfer coefficients. A cone is placed in an air stream with its axis parallel to the direction of the flow. The cone has a surface area and mass of $0·32 \times 10^{-2}$ m² and $0·124$ kg respectively and the material from which it is made has a specific heat of $0·4 \times 10^3$ J/kg K. In an experiment, the cone is heated to 361 K and then allowed to cool in the air stream which is at 296·9 K. The results of the experiment are as follows:

Time	(s)	0	150	300	450	600	750
Temperature	(K)	361	330	313·6	305·5	301·4	299·3

Calculate the heat transfer coefficient at the surface of the cone at this velocity.

(68 W/m² K)

Problem 4-6: Two qualities, A and B, of insulating material are available for the insulation of a 120 mm diameter steam distribution system. Insulation A has a thermal conductivity of 0·08 W/m K and costs £9·50/m³. Insulation B has a thermal conductivity 0·05 W/m K and costs £14·00/m³. The heat transfer coefficient at the surface of the insulation, which may be assumed independent of surface diameter and temperature, is estimated to be 12 W/m²K. If 0·55/m length may be spent on insulation, determine which material should be used. Calculate its thickness, and the heat loss per unit length, unit overall temperature difference, when this insulation is installed.

(B, 57·6 mm, 0·4 W/m K)

Problem 4-7: A spherical vessel, 1·2 m diameter is to be used for the storage of liquid nitrogen at 77 K in an area where the relative humidity of the air surrounding the sphere will be 80% at 305 K. What thickness of insulation, k = 0·02 W/m K, will be necessary to prevent condensation on the insulation surface? The heat transfer coefficient at the insulation surface may be taken as 12 W/m²K.

(83 mm)

Problem 4-8: What do you understand by the term fin efficiency. Show that the fin efficiency η for a fin of constant cross sectional area A, length L with zero tip loss is given by

$$\eta = \frac{1}{mL} \tanh mL \quad \text{where } m = \left(\frac{hP}{kA}\right)^{\frac{1}{2}}$$

h is the heat transfer coefficient at the fin surface, k the thermal conductivity of the fin material, P the fin perimeter.

A surface is provided with longitudinal fins, 25·4 mm high and 2·54 mm thick. The pitch of the fins is 7·62 mm. If h and k are 146·5 W/m²K and 208 W/m K show that by halving the fin thickness and doubling the number of fins, heat dissipation from the surface is increased by about 80%.

Problem 4-9: The temperature of superheated steam, flowing in a large duct is to be measured by means of a thermocouple attached to the end of a steel tube 6 mm outside diameter with 1·6 mm wall thickness, 150 mm long mounted normal to the duct wall. If the thermocouple indicates a steam temperature of 503 K estimate the actual steam temperature if the duct wall temperature is 308 K. The heat transfer coefficient between the tube and the steam may be assumed constant at 30 W/m²K. Heat flow into the exposed end of tube may be neglected. The thermal conductivity of steel is 42 W/m K.

(512·9 K)

Problem 4-10: The side panels of a steel duct are to be braced by steel tubes, 40 mm outside diameter, spanning the duct width which is 1·25 m. During operation the duct will pass reactor exhaust gases at a temperature of 1023 K and the maximum heat transfer coefficient between this gas and the bracing tubes is estimated as 2·5 W/m²K. If the duct walls are maintained at 473 K, what is the minimum thickness of the bracing tube wall required in order that the maximum permissible metal temperature of 823 K shall not be exceeded? Radial temperature gradients in the tube wall are to be neglected. Thermal conductivity of steel is 42 W/m K.

(2·2 mm)

Problem 4-11: If the setting of concrete is accompanied by the diffuse evolution of heat at the rate of 50 W/m³, determine the maximum temperature in a concrete wall 1 m thick during its setting period. It may be assumed that steady state conditions prevail through the setting period and that the surface temperature of the wall is maintained at 281 K. Thermal conductivity of concrete 0·8 W/m K.

(288·8 K)

Problem 4-12:   Refer to Example 4-4 and calculate the number of sheets in the stack if one of the plate surfaces is maintained at a constant temperature of 333 K, whilst the other remains at 343 K. If this stack were being processed and the surface initially at 333 K suddenly became starved of cooling medium such that no heat could be dissipated through that surface, how much of the stack would be damaged?

(160, 91%)

Problem 4-13:   A vertical furnace wall has an inner lining of firebrick 0·2 m thick which is covered by a layer of magnesia 0·3 m thick. If the inner surface temperature of the firebrick is 800 K and the ambient air temperature T_a is 293 K calculate:

(a) the exposed surface temperature of the magnesia (T_s K).
(b) the heat loss through the wall (W/m²), and
(c) the firebrick/magnesia interface temperature.

Thermal conductivity of fire brick 0·31 W/m K.
Thermal conductivity of magnesia 0·14 W/m K.
The heat transfer coefficient at the exposed surface of the magnesia is given by $h = 1·96 (T_s - T_a)^{0·25}$ where h is in W/m²K and $(T_s - T_a)$ is in K.

(328·5 K, 169 W/m², 691 K)

117

5. Basic Mass Transfer

5.1 Introduction

The aim of this chapter is to provide a general introduction to the mechanism of mass transfer. In diffusional operations, mass transfer occurs as a result of a concentration difference, with the diffusing substance moving from a position of high concentration to one of low concentration. In the evaporation of a liquid for example, molecules of the vapour diffuse through those of the surroundings and are carried away. Diffusion may occur through a substance which is itself diffusing in the opposite direction. Equation 5-1 represents the general case:

$$\mathrm{d}N_a/\mathrm{d}t = KA\Delta C \tag{5-1}$$

where N_a is the quantity of material a transferred t, the time, K a constant, A the cross-sectional area, and ΔC the concentration gradient.

This equation is analogous to the equation governing the rate of heat transfer by conduction and convection, and in a similar way the constant K is dependent upon the properties of the fluid as well as the dynamic and geometric properties of the system.

This chapter is concerned with aspects and applications of eqn. 5-1 and how the resulting constants and mass transfer coefficients may be determined.

5.2 Molecular diffusion of gases

Molecular diffusion may be defined as the transport of matter on a molecular scale through a stagnant fluid or, if the fluid is in laminar flow, in a direction perpendicular to the flowing streamlines. (Diffusion through a turbulent fluid occurs by eddy diffusion).

By considering the molecular diffusion process in gases and relating the concentration gradient to the partial pressures of the components, the following general equation may be derived

$$-dP_a = (RT/D_{ab}P)\,(N_aP - N_aP_a - N_bP_b)\,dx \qquad (5\text{-}2)$$

where P is the total pressure (kN/m^2) R the gas constant $(8\cdot314$ kJ/kmol K) T the temperature (K), D_{ab} the diffusion coefficient (m^2/s) x the distance in the direction of diffusion (m) P_a, P_b the partial pressures of components a and b (kN/m^2).

Two particular cases will be considered, but the diffusivity D is worthy of further attention at this stage.

5.2.1 Diffusivity of gases and vapours

The diffusivity or diffusion coefficient D depends upon the temperature, pressure and nature of the components of the system. Experimental data are available for commonly encountered systems and an example of the calculation of D from typical experimental data follows.

Example 5-1:   A small diameter tube closed at one end was filled with acetone to within 18 mm of the top and maintained at 290 K with a gentle stream of air blowing across the top (Fig. 5-1). After 15 ks, the

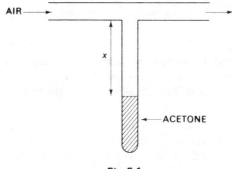

Fig. 5-1

liquid level had fallen to 27·5 mm. The vapour pressure of acetone was 21·95 kN/m² and atmospheric pressure was 99·75 kN/m². Calculate the diffusivity of acetone in air.

Solution   The system is illustrated in Fig. 5-1.

It can be shown that the rate of evaporation/unit area is

$$N_A = \frac{DP}{RTx} \cdot \ln \frac{P_{B_2}}{P_{B_1}} = \frac{\rho_L}{M} \cdot \frac{dx}{dt} \tag{5-3}$$

where R is 8·314 kJ/kmol K, T is 290 K, P is 99.75 kN/m², ρ_L is 790 kg/m³, and M = 58 kg/kmol.

From eqn. 5-3

$$dt \left(\frac{DP}{RT} \cdot \ln \frac{P_{B_2}}{P_{B_1}} \right) \frac{M}{\rho_L} = x \, dx$$

therefore

$$\left(\frac{DP}{RT} \ln \frac{P_{B_1}}{P_{B_2}} \cdot \frac{M}{\rho_L} \right) \cdot \int_0^t dt = \int_{x_1}^{x_2} x \, dx$$

giving

$$\frac{DP}{RT} \ln \frac{P_{B_1}}{P_{B_2}} \frac{M}{\rho_L} \cdot t = \frac{1}{2}(x_2^2 - x_1^2) \tag{5-4}$$

From eqn. 5-4,

$$D = \frac{RT\rho_L(x_2^2 - x_1^2)}{2 \cdot P \cdot \ln \left(\frac{P_{B_1}}{P_{B_2}} \right) Mt}$$

$$= \frac{8·314 \times 290 \times 790 \, (0·0275^2 - 0·018^2)}{2 \times 99·75 \times \ln \, (99·75/(99·75 - 21·95)) \times 58 \times 15 \, 000}$$

$$= 1·84 \times 10^{-5} \text{ m}^2/\text{s}$$

Where literature values do not exist and experimental work is not practicable, methods of estimating the diffusivity for both gases and liquids are available and these are illustrated in the following examples.

Example 5-2:   Estimate the value of the diffusion coefficient of benzene vapour in air at a temperature of 273 K and a pressure of 101·3 kN/m². The critical temperature and critical volume of air may be taken as 132·3 K and 0·002 99 m³/kg respectively.

Solution For two components a and b the diffusion coefficient D_V may be estimated by a method due to Chen and Othmer from the equation

$$(D_V)_{ab} = (D_V)_{ba}$$

$$= \frac{6 \cdot 05 \times 10^{-4} \cdot T^{1 \cdot 81} \cdot [1/(MW)_a + 1/(MW)_b]^{0 \cdot 5}}{P_T [(T_c)_a (T_c)_b]^{0 \cdot 1405} \cdot \{[(MW)_a (v_c)_a]^{0 \cdot 4} + [(MW)_b (v_c)_b]^{0 \cdot 4}\}^2}$$

(5-5)

where T is temperature (K), MW the molecular weight (kg/kmol), P_T the total pressure (N/m²), T_c the critical temperature (K), v_c the critical volume (m³/kg). Subscripts a and b refer to components a and b.

Where critical properties are not available, they may be estimated from

$$T_B/T_c = 0 \cdot 567 + \Sigma\phi(T) - [\Sigma\phi(T)]^2$$ (5-6)

where T_B = normal boiling temperature (354 K for benzene)

$$P_c = 101\,325(MW)/[0 \cdot 340 + \Sigma\phi(P)]^2$$ (5-7)

and

$$v_c = 1/\rho_c = [0 \cdot 040 + \Sigma\phi(V)]/MW$$ (5-8)

$\phi(T)$, $\phi(P)$, $\phi(V)$ refer to the structural values of the group contributions (see Table 5-1).

Benzene has a structure of 6 x =CH (ring) and for this structure

$\phi(V)$, $\phi(T)$ are $0 \cdot 037$ m³/kmol and $0 \cdot 011$ respectively.

Thus the critical volume

$$v_c = [0 \cdot 040 + (6 \times 0 \cdot 037)]/78 = 0 \cdot 003\,358 \text{ m}^3/\text{kg}$$

The critical temperature

$$T_c = \frac{354}{[0 \cdot 567 + (6 \times 0 \cdot 011) - (6 \times 0 \cdot 011)^2]} = 562 \text{ K}.$$

If subscripts a and b refer to air and benzene respectively, the following substitution may be made into eqn. 5-5

$$D_V = \frac{6 \cdot 05 \times 10^{-4} (273)^{1 \cdot 81} [1/29 + 1/78]^{0 \cdot 5}}{101\,300 [132 \cdot 3 \times 562]^{0 \cdot 1405} \cdot \{[29 \times 0 \cdot 002\,99]^{0 \cdot 4} + [78 \times 0 \cdot 003\,358]^{0 \cdot 4}\}^2}$$

$$= 7 \cdot 54 \times 10^{-6} \text{ m}^2/\text{s}$$

It should be noted that the product of diffusivity and pressure is a constant for pressures in the range $0 \cdot 3$ to 5000 kN/m². An example of the diffusivity of one component through a multicomponent mixture is included in Example 5-5.

5.2.2 Equimolecular counter-diffusion

If in eqn. 5-2, N_a and N_b are constant and $N_a = -N_b$, the condition of equimolecular counter-diffusion is said to exist. Component a is shown in Fig. 5-2 to be diffusing in the direction of its drop in concentration and b is diffusing in the opposite direction. Distillation operations form good examples of this process and the concentrations at any point in the gas mixture remain constant with time.

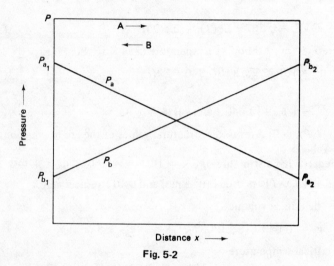

Fig. 5-2

In eqn. 5-2, if $N_a = -N_b$ and D_{ab} may be assumed constant, then

$$-\int_{P_{a_1}}^{P_{a_2}} dP_a = \frac{RTN_a}{D_{ab}} \int_{x_1}^{x_2} dx \tag{5-9}$$

Integrating and putting $x_2 - x_1 = x$ gives

$$N_a = \frac{D_{ab}}{RTx} (P_{a_1} - P_{a_2}) \tag{5-10}$$

Equation 5-10 is known as Fick's Law and its application is shown in Example 5-3.

Table 5-1   Structural Values for the Estimation of Critical Properties

Structural value	$\phi(T)$	$\phi(P)$ (kg/kmol$(101\cdot3$ kN/m^3))$^{0\cdot5}$	$\phi(V)$ m^3/kmol
Structure			
Non-ring:			
—CH$_3$	0·020	0·227	0·055
—CH$_2$	0·020	0·227	0·055
—CH	0·012	0·210	0·051
—C—	0·000	0·210	0·041
=CH$_2$	0·018	0·198	0·045
=CH	0·018	0·198	0·045
=C—	0·000	0·198	0·036
=C=	0·000	0·198	0·036
≡CH	0·005	0·153	(0·036)
≡C—	0·005	0·153	(0·036)
Ring:			
—CH$_2$—	0·013	0·184	0·0445
—CH	0·012	0·192	0·046
—C—	(−0·007)	(0·154)	(0·031)
=CH—	0·011	0·154	0·037
=C—	0·011	0·154	0·036
=C=	0·011	0·154	0·036
Halogen:			
—F	0·018	0·224	0·018
—Cl	0·017	0·320	0·049
—Br	0·010	(0·500)	(0·070)
—I	0·012	(0·830)	(0·095)

Table 5-1 (*continued*)

structural value	$\phi(T)$	$\phi(P)$ (kg/kmol(101.3 kN/m³))$^{0.5}$	$\phi(V)$ (m³/kmol)
Oxygen:			
—OH(alcohols)	0·082	0·060	(0·018)
—OH(phenols)	0·031	(−0·020)	(0·003)
—O—(nonring)	0·021	0·160	0·020
—O—(ring)	(0·014)	(0·120)	(0·008)
—C=O(nonring)	0·040	0·290	0·060
—C=O(ring)	(0·033)	(0·200)	(0·050)
H—C=O(aldehyde)	0·048	0·330	0·073
—COOH(acid)	0·085	(0·400)	0·080
—COO—(ester)	0·047	0·470	0·080
=O(except as above)	(0·020)	(0·120)	(0·011)
Nitrogen:			
—NH₂	0·031	0·095	0·028
—NH(nonring)	0·031	0·135	(0·037)
—NH(ring)	(0·024)	(0·090)	(0·027)
—N—(nonring)	0·014)	0·170	(0·042)
—N—(ring)	(0·007)	(0·130)	(0·032)
—CN	(0·060)	(0·360)	(0·080)
—NO₂	(0·055)	(0·420)	(0·078)
Sulphur:			
—SH	0·015	0·270	0·055
—S—(nonring)	0·015	0·270	0·055
—S—(ring)	(0·008)	(0·240)	(0·045)
=S	(0·003)	(0·240)	(0·047)

Values in parenthesis are based on too few experimental values to be reliable. Values for hydrogen are included in the group value and bonds shown as unconnected are connected to an atom other than hydrogen.

Example 5-3: At a point in an atmospheric pressure distillation column containing a benzene-toluene mixture, the temperature is 372 K and the liquid and vapour phases contain 30 mol % and 40 mol % benzene respectively. The vapour pressure of toluene is 72·0

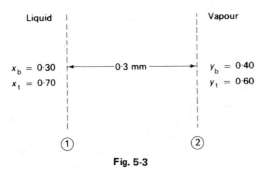

Fig. 5-3

kN/m^2 at this temperature and its diffusivity may be taken as 5.4×10^{-6} m^2/s. If the resistance to mass transfer lies in a film 0.3 mm thick, calculate the rate of interchange of benzene and toluene between the liquid and vapour phases.

Solution   Using subscripts b and t for benzene and toluene respectively, the situation existing is shown in Fig. 5-3.

Providing the heat losses are negligible from the system and that the molal latent heats are equal, it may be assumed that $N_t = N_b$.

At plane 1, the partial pressure of toluene is obtained from Raoult's Law as

$$P_{t_1} = 0.7 \times 72.0 = 50.4 \text{ kN/m}^2$$

and

$$P_{t_2} = 0.6 \times 101.3 = 60.8 \text{ kN/m}^2$$

Hence for toluene

$$N_t = \frac{5.4 \times 10^{-6}}{8.314 \times 372 \times 0.3 \times 10^{-3}} (60.8 - 50.4)$$
$$= 60.5 \times 10^{-6} \text{ kmol/s m}^2$$

This is the rate of transfer of toluene from vapour to liquid. Benzene will pass in the opposite direction at the same rate.

5.2.3 Diffusion through a stagnant layer

The situation which exists when one material diffuses through a stagnant layer of another is shown in Fig. 5-4. Component a diffuses by

125

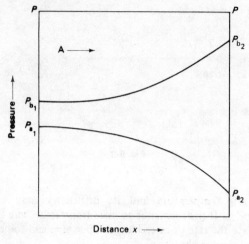

Fig. 5-4

virtue of its driving force $(P_{a_1} - P_{a_2})$ while b does not diffuse but maintains its concentration gradient by intermolecular friction between a and b. In eqn. 5-12, the hindering effect is included by the logarithmic mean partial pressure of b, P_{bm}.

The corresponding equation to that for equimolecular counter diffusion derived from eqn. 5-2 with $N_b = 0$ and N_a constant is

$$N_a = \frac{D_{ab}}{RTx} \cdot \frac{P}{P_{bm}} (P_{a_1} - P_{a_2}) \qquad (5\text{-}11)$$

where P is total pressure (kN/m^2) and P_{bm} the log mean partial pressure of b (kN/m^2).

Example 5-4: A layer of benzene 1 mm deep lies at the bottom of an open tank 5 m in diameter. The tank temperature is 295 K and the diffusivity of benzene in air is $8\cdot0 \times 10^{-6}$ m^2/s at this temperature. If the vapour pressure of benzene in the tank is $13\cdot3$ kN/m^2 and diffusion may be assumed to take place through a stagnant air film 3 mm thick, how long will it take for the benzene to evaporate? The density of benzene is 880 kg/m^3.

Solution In this case $x = 3$ mm $= 0\cdot003$ m, $P_{a_1} = 13\cdot3\,kN/m^2$, $P_{a_2} = 0$, $P_{b_1} = (101\cdot3 - 13\cdot3) = 88\cdot0\,kN/m^2$, $P_{b_2} = 101\cdot3\,kN/m^2$, and

$$P_{bm} = \frac{101 \cdot 3 - 88 \cdot 0}{\ln \left(\dfrac{101 \cdot 3}{88 \cdot 0}\right)} = 93 \cdot 27 \text{ kN/m}^2.$$

Substituting in eqn. 5-11 gives the rate of diffusion of benzene

$$N_a = \frac{8 \cdot 0 \times 10^{-6}}{8 \cdot 314 \times 295 \times 0 \cdot 003} \cdot \frac{101 \cdot 3}{93 \cdot 27} (13 \cdot 3 - 0) = 15 \cdot 71 \times 10^{-6} \text{ kmol/s m}^2$$

The area of the tank is

$$\frac{\pi}{4} \times 5^2 = 19 \cdot 64 \text{ m}^2$$

Hence the rate of diffusion is

$$19 \cdot 64 \times 15 \cdot 71 \times 10^{-6} = 3 \cdot 09 \times 10^{-4} \text{ kmol/s}$$

The mass of benzene to be evaporated is given by

$$19 \cdot 64 \times 0 \cdot 001 \times 880 = 17 \cdot 28 \text{ kg} = 0 \cdot 222 \text{ kmol}$$

Hence the time to evaporate is

$$0 \cdot 222/(3 \cdot 09 \times 10^{-4}) = 718 \text{ s}$$

Example 5-5: (a) Nitrogen is diffusing under steady state conditions through C_4H_{10} at 298 K and a total pressure of 100 kN/m². The partial pressures of nitrogen at two planes 0·01 m apart are 13·3 and 6·67 kN/m² respectively. If the diffusivity of nitrogen through C_4H_{10} is $9·6 \times 10^{-6}$ m²/s, calculate the rate of diffusion of nitrogen across the two planes.

(b) If instead of pure C_4H_{10}, a mixture of 20 vol % of C_2H_6, 30% C_2H_4 and 50% C_4H_{10} is the non-diffusing gas, what will the rate of diffusion of nitrogen be if the other conditions remain the same? The diffusivities of nitrogen through C_2H_6 and C_2H_4 may be taken as $14·8 \times 10^{-6}$ and $16·3 \times 10^{-6}$ m²/s respectively.

Solution (a) if nitrogen is regarded as component a, the required value of N_a is given by

$$N_a = \frac{D_a}{RTx} \cdot \frac{P}{P_{bm}} \cdot (P_{a_1} - P_{a_2}) \tag{5-11}$$

In this case, $P_{a_1} = 13 \cdot 3 \text{ kN/m}^2$, $P_{a_2} = 6 \cdot 67 \text{ kN/m}^2$, $P_{b_1} = 86 \cdot 7 \text{ kN/m}^2$, $P_{b_2} = 93 \cdot 3 \text{ kN/m}^2$.

Hence

$$P_{bm} = 89 \cdot 67 \text{ kN/m}^2$$

and

$$N_a = \frac{9 \cdot 6 \times 10^{-6}}{8 \cdot 314 \times 298 \times 0 \cdot 01} \cdot \frac{100}{89 \cdot 67} (13 \cdot 3 - 6 \cdot 67)$$

$$= 2 \cdot 88 \times 10^{-6} \text{ kmol/s m}^2$$

(b) For steady state diffusion through a non-diffusing multicomponent mixture of constant composition, an effective diffusivity may be calculated from

$$D_a' = \frac{1}{(y_b/D_{ab}) + (y_c/D_{ac}) + (y_d/D_{ad})} + \ldots \qquad (5\text{-}12)$$

where y_b, y_c, y_d .. = mole fraction compositions of the mixture on an a-free basis

D_{ab}, D_{ac}, D_{ad} .. diffusivities of a through b, c, d . . .

In addition, the logarithmic mean partial pressure of P_{bm} is replaced by the mean partial pressure of the non-diffusing mixture, P_{im}.
In this example

$$D_a' = \frac{1}{\dfrac{0 \cdot 20}{14 \cdot 8 \times 10^{-6}} + \dfrac{0 \cdot 30}{16 \cdot 3 \times 10^{-6}} + \dfrac{0 \cdot 50}{9 \cdot 6 \times 10^{-6}}} = 11 \cdot 9 \times 10^{-6} \text{ m}^2/\text{s}$$

Also in this case

$$P_{im} = P_{bm} \text{ from part (a)} = 89 \cdot 67 \text{ kN/m}^2$$

Hence

$$N_a' = \frac{11 \cdot 9 \times 10^{-6}}{8 \cdot 314 \times 298 \times 0 \cdot 01} \times \frac{100}{89 \cdot 67} (13 \cdot 3 - 6 \cdot 67)$$

$$= 3 \cdot 55 \times 10^{-6} \text{ kmol/s m}^2$$

5.3 Molecular diffusion in liquids

By an analogous treatment to that for diffusion in gases, expressions may be derived for the rate of diffusion in liquids. A greater degree of uncertainty exists in the resulting expressions, however, since theories on the physical property variations with concentration in liquids are less well developed than with gases.

5.3.1 Diffusivity of liquids

Estimation methods for diffusivity in liquids in the absence of experimental data are complex, but for non-electrolytes in dilute solution the following method may be used.

Example 5-6:   Estimate the diffusivity of benzene in dilute aqueous solution at 298 K. The viscosity of water at 298 K may be taken as 8.94×10^{-4} Ns/m^2.

Solution   The diffusivity is estimated from a method due to Wilke which states that

$$F = T/D_{ab}\mu_L$$

where F is the value obtained from Fig. 5-5, T the temperature (K), D_{ab} the diffusivity of solute a through solvent b (m^2/s), and μ the viscosity of solvent (N s/m^2)

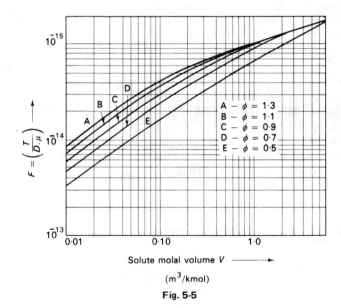

Fig. 5-5

The parameter ϕ in Fig. 5-5 is equal to the ratio of the value of F in the solvent to the value of F for diffusion in water at constant molal volume. For water, methanol, and benzene the values of ϕ are $1.0, 0.82$

129

and 0·72 respectively and if experimental data are lacking for other solvents, a value of 0·90 may be assumed.

The solute molal volume in Fig. 5-5 may be obtained from Table 5-2 or calculated from the atomic volumes to give V. Using this value of V and the appropriate value of ϕ, the value of the function F may be found, from which the diffusivity may be calculated.

Table 5-2   Atomic and molecular volumes

Atomic volume (m³/kmol)		Molecular volume (m³/kmol)	
Carbon	0·0148	H_2	0·0143
Hydrogen	0·0037	O_2	0·0256
Chlorine	0·0246	N_2	0·0312
Bromine	0·0270	Air	0·0299
Iodine	0·0370	CO	0·0307
Sulphur	0·0256	CO_2	0·0340
Nitrogen	0·0156	SO_2	0·0448
Nitrogen in primary amines	0·0105	NO	0·0236
Nitrogen in secondary amines	0·0102	N_2O	0·0364
Oxygen	0·0074	NH_3	0·0258
Oxygen in methyl esters	0·0091	H_2O	0·0189
Oxygen in higher esters	0·0110	H_2S	0·0329
Oxygen in acids	0·0120	COS	0·0515
Oxygen in methyl ethers	0·0099	Cl_2	0·0484
Oxygen in higher ethers	0·0110	Br_2	0·0532
Benzene ring: subtract	0·0150	I_2	0·0715
Naphthalene ring: subtract	0·0300		

Using Table 5-2 to calculate the molal volume of the solute benzene

$$V_a = (6 \times 0·0148) + (6 \times 0·0037) - (0·0015) = 0·1095 \text{ m}^3/\text{kmol}$$

From Fig. 5-5, when $V = 0·096$ and $\phi = 1$ (for water)

$$F = 3·24 \times 10^{-14} = T/D_{ab}\mu_L$$

Hence

$$D_{ab} = 298/(3·24 \times 10^{-14} \times 8·94 \times 10^{-4}) = 1·03 \times 10^{-9} \text{ m}^2/\text{s}$$

(This compares well with an experimental value of $1·0 \times 10^{-5}$ cm²/s quoted in Perry.)

It has been found that F is virtually independent of temperature, so that, providing the variation of solvent viscosity with temperature is known, the diffusivity may be calculated at different temperatures.

5.3.2 Steady-state equimolal counter-diffusion

It can be shown that Fick's Law applies to this situation and it is written as

$$N_a = \frac{D_{ab}}{Z} \cdot (C_{a_1} - C_{a_2}) \tag{5-14}$$

where C_{a_1} and C_{a_2} are the concentrations of component a at two planes distance Z apart.

If $C_a + C_b = C$, then $C_a = x_a C$ and eqn. 5-14 may be written in terms of mole fractions x as

$$N_a = \frac{D_{ab} \cdot C}{Z} (x_{a_1} - x_{a_2}) \tag{5-15}$$

5.3.3 Diffusion through a stagnant (non-diffusing) layer

As for the case of gases, eqns. 5-14 and 5-15 are modified by a factor C_{bm} or x_{bm} to give

$$N_a = \frac{D_{ab} \cdot C(C_{a_1} - C_{a_1})}{C_{bm} \cdot Z} = \frac{D_{ab} \cdot C(x_{a_1} - x_{a_2})}{x_{bm} \cdot Z} \tag{5-16}$$

5.4 Mass transfer between gas and liquid phases

The transfer of a gas to a liquid may be regarded as taking place in three stages

     (i) diffusion from the gas to the liquid surface,
    (ii) solution in the liquid, and
   (iii) diffusion from the liquid surface to the bulk of the liquid.

This process is shown in Fig. 5-6. If the values of P_i and C_i, the partial pressure and concentration at the interface were known, it would be possible to use the equations for diffusion to calculate the rate of mass transfer. In practice, the distances over which the diffusion takes place are not normally known as they depend upon the dynamic characteristics of the gas and liquid streams.

However, if use is made of individual and overall mass transfer coefficients defined below, rates of mass transfer can be obtained without knowledge of these film thicknesses.

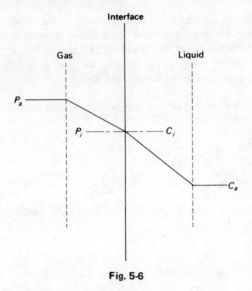

Fig. 5-6

Individual mass transfer coefficients are defined as

$$k_G = N_a/(P_a - P_i) \tag{5-17}$$

$$k_L = N_a/(C_i - C_a) \tag{5-18}$$

where N_a is the rate of transfer (kmol/s m^2), P_a, P_i the partial pressure of component a in gas and at interface (kN/m^2), C_a, C_i the concentration of solute a in liquid and at interface (kmol/m^3), k_G the gas film coefficient (kmol/s m^2 (kN/m^2)), and k_L the liquid film coefficient (kmol/s m^2 (kmol/m^3)). Overall coefficients are normally defined on the basis of the gas phase by K_G or the liquid phase by K_L given by the equations

$$K_G = N_a/(P_a - P^*) \tag{5-19}$$

$$K_L = N_a/(C^* - C_a) \tag{5-20}$$

where P^* is the partial pressure in equilibrium with C_a and C^* is the concentration in equilibrium with P_a. If the solution obeys Henry's Law

$$P = HC$$

where P and C are the partial pressure and concentration at equilibrium and H is a constant, then

$$P_a = HC^* \tag{5-21}$$

$$P^* = HC_a \qquad (5\text{-}22)$$

$$P_i = HC_i \qquad (5\text{-}23)$$

and

$$\frac{1}{K_G} = \frac{P_a - P^*}{N_a} = \frac{P_a - P_i}{N_a} + \frac{H(C_i - C_a)}{N_a} \qquad (5\text{-}24)$$

$$= \frac{1}{k_G} + \frac{H}{k_L} \qquad (5\text{-}25)$$

and

$$\frac{1}{K_L} = \frac{C^* - C_a}{N_a} = \frac{P_a - P_i}{H \cdot N_a} + \frac{C_i - C_a}{N_a} \qquad (5\text{-}26)$$

$$= \frac{1}{H \cdot k_G} + \frac{1}{k_L} \qquad (5\text{-}27)$$

Thus, for very soluble gases, e.g. ammonia in water, where H is very small, $K_G \simeq k_G$. For very low solubility gases, eg. oxygen in water, H is large and $K_L \simeq k_L$. These two situations give rise to the expressions *gas film control* and *liquid film control* respectively.

Example 5-7: Using the preceding relationships, demonstrate graphically how the driving forces are related in the following cases:

  (a) Where the individual mass transfer coefficients k_L and k_G are known.
  (b) Where P_a and C_a are measured experimentally.
  (c) As for (b) but where the equilibrium line is straight with a small value of the slope, m.
  (d) As for (c) but where the value of m is large.

Solution   (a) Equations 5-17 and 5-18 may be rewritten as

$$N_a = k_G(P_a - P_i) = k_L(C_a - C_i) \qquad (5\text{-}28)$$

or the driving force ratio

$$(P_a - P_i)/(C_a - C_i) = - k_L/k_G \qquad (5\text{-}29)$$

The equilibrium curve has the equation $P_i = f(C_i)$ and the system is illustrated in Fig. 5-7. The point M represents the equilibrium between a point P_i and C_i. According to eqn. 5-29, a line of slope $-k_L/k_G$ (which is known) drawn through M to P will depict conditions at a point in the

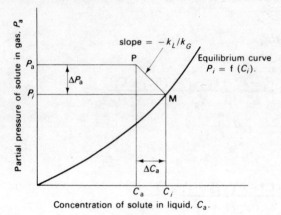

Fig. 5-7

actual operation at P_a and C_a. The driving forces ΔP_a and ΔC_a are then derived graphically to enable N_a to be found for this particular point in the system

(b) As in part (a) eqns. 5-17 to 5-20 may be combined to give

$$N_a = k_G(P_a - P_i) = K_G(P_a - P^*) = K_G \Delta P_{oa} \tag{5-28}$$

and

$$N_a = k_L(C_i - C_a) = K_L(C^* - C_a) = K_L \Delta C_{oa} \tag{5-29}$$

where ΔP_{oa}, ΔC_{oa} refer to the overall driving forces.

Sampling and analysis in the actual system provides values of P_a and C_a from the bulk of the two phases. (It is physically impossible to obtain similar values of P_i and C_i owing to the very small film thicknesses.) Thus the point P is located on Fig. 5-8. The overall driving forces may be indicated as ΔP_{oa} and ΔC_{oa} as the vertical and horizontal distances from P to the equilibrium curve at A and B as shown in Fig. 5-8.

If the individual coefficients k_L and k_G are known, the point M may be located as shown in the example of part (a). This analysis assumes that the equilibrium curve is straight over the particular concentration range and by considering the geometry of Fig. 5-8 we can write

$$\Delta P_{oa} = (k_L/k_G)\Delta C_a + m\Delta C_a = \Delta C_a((k_L/k_G) + m) \tag{5-30}$$

Also

$$\Delta P_{oa} K_G = N_a = K_G((k_L/k_G) + m)\Delta C_a = k_L \Delta C_a \tag{5-31}$$

134

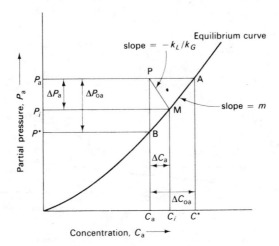

Fig. 5-8

then

$$K_G((k_L/k_G) + m) = k_L \qquad (5\text{-}32)$$

and

$$1/K_G = 1/k_G + m/k_L \qquad (5\text{-}25)$$

Similarly

$$1/K_L = 1/mk_G + 1/k_L \qquad (5\text{-}26)$$

which are the relationships obtained earlier.

(c) The situation where m is small is that in which the gas film is controlling and $1/K_G \simeq 1/k_G$ (solute a is very soluble in the liquid) or

$$\Delta P_{oa} \simeq \Delta P_a$$

From Fig. 5-9 it will be seen that a fairly large change in k_L, altering the slope of the line PM, will have little effect on the value of K_G. In this case, in order to increase the value of K_G it would be desirable to decrease the gas phase resistance by maximum agitation of the gas. This may be achieved, for example, by allowing the liquid to become dispersed as droplets in the gas.

(d) The converse of (c) when m is very large is shown in Fig. 5-10 where $1/K_L \simeq 1/K_L$, and

$$\Delta C_{oa} \simeq \Delta C_a$$

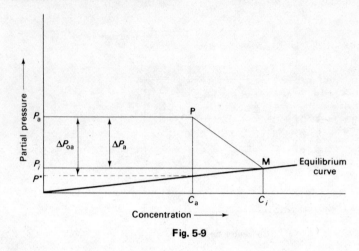

Fig. 5-9

A change in k_G produces relatively little change in K_L and for this reason the situation is known as *liquid film controlled*.

Wetted wall columns provide a convenient means of obtaining values of the individual mass transfer coefficients from which the overall coefficients required for design purposes may be calculated. The following examples serve to illustrate the techniques involved.

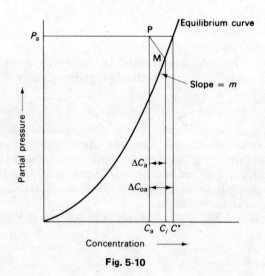

Fig. 5-10

Example 5-8: Calculate the gas film coefficient k_G for the absorption of sulphur dioxide from a dilute mixture with air in a wetted wall column using the experimental data below.

Inside diameter of column, d = 25 mm
Gas velocity, v = 2·2 m/s
Gas temperature, T = 293 K
Gas viscosity, μ = 1·78 x 10^{-5} N s/m^2
Gas density, ρ = 1·22 kg/m^3
Diffusivity, D = 12·2 x 10^{-6} m^2/s
Gas constant, R = 8·314 kJ/kmol K

Solution   Use is made of Gilliland and Sherwood's correlation, that is

$$d/x = 0·023 \, Re^{0·83} \cdot Sc^{0·44} \tag{5-33}$$

where Re is the Reynolds Number (= $\rho v d/\mu$), Sc the Schmidt Number (= $\mu/\rho D$), x the effective film thickness (mm), and d the column diameter (mm).
In this example

$$Re = 1·22 \times 2·2 \times 0·025/1·78 \times 10^{-5} = 3770$$
$$Sc = 1·78 \times 10^{-5}/1·22 \times 12·2 \times 10^{-6} = 1·2$$

Hence

$$d/x = 0·023(3770)^{0·83} (1·2)^{0·44} = 22·9$$

and

$$x = 25·0/22·9 = 1·09 \text{ mm}$$

Now

$$N_a = k_G(P_i - P_a) \tag{5-17}$$

where P_a, P_i are the partial pressures of solute a in gas and at the interface (kN/m^2).

But

$$N_a = \frac{D}{RTx} \cdot \frac{P}{P_{bm}} (P_i - P_a) \tag{5-11}$$

hence

$$k_G = \frac{D}{RTx} \cdot \frac{P}{P_{bm}} \tag{5-34}$$

For a dilute gas mixture, P/P_{bm} is approximately equal to unity and

$$k_G = \frac{DP}{RTx} = \frac{12 \cdot 2 \times 10^{-6} \; 101 \cdot 3}{8 \cdot 314 \times 293 \times 0 \cdot 00109} \times 1$$

$$= 4 \cdot 65 \times 10^{-4} \; \text{kmol/s m}^2 \, (\text{kN/m}^2)$$

Example 5-9: The following data were obtained from a wetted-wall column employing a constant liquid flow rate.

molar gas flow rate G_m (kmol/s)	0·01	0·02	0·04	0·06	0·08	0·10
overall mass transfer K_G coefficient (kmol/m² s (kN/m²) × 10⁶)	50·8	67·8	84·0	91·7	93·5	100

The relationship between the equilibrium vapour pressure P_A (kN/m²) and the molar concentration in the liquid phase C_A (kmol/m³) is given by

$$P_A = 20C_A$$

For a gas flow rate of 0·05 kmol/s calculate the overall and individual mass transfer coefficients.

Solution
It has been shown that

$$1/K_G = (1/k_G) + (H/k_L) \tag{5-25}$$

where k_G, k_L are individual mass transfer coefficients, and H is Henry's constant = 20 (kN/m²)/(kmol/m³).

It has been shown experimentally that k_G is proportional to (gas velocity)$^{0 \cdot 8}$ and hence $k_G = \alpha(G_m)^{0 \cdot 8}$. A plot of $1/k_G$ vs $1/G_m{}^{0 \cdot 8}$ will yield a straight line plot from which k_L and α may be obtained. This is shown in Fig. 5-11.
From Fig. 5-11, the intercept

$$H/k_L = 8150$$

therefore

$$k_L = 20/8150 = 0 \cdot 0025 \; \text{m/s}$$

Similarly, the slope

$$\frac{1}{\alpha} = 288 \cdot 8$$

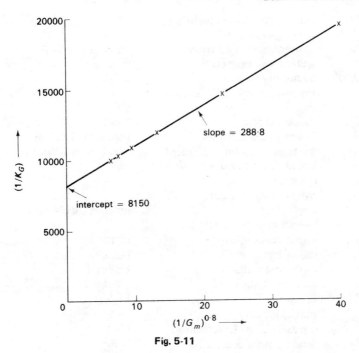

Fig. 5-11

hence

$$\alpha = 0 \cdot 00346$$

Now, at $G_m = 0 \cdot 05$, $G_m^{0 \cdot 8} = 0 \cdot 092$, hence

$$k_G = \alpha G_m^{0 \cdot 8} = 0 \cdot 003\ 46 \times 0 \cdot 092 = 0 \cdot 000\ 32 \text{ kmol/m}^2\text{s(kN/m}^2)$$

and

$$1/K_G = (1/0 \cdot 000\ 32) + 8150 = 11\ 275$$

therefore

$$K_G = 88 \cdot 6 \times 10^{-6} \text{ kmol/m}^2 \text{ s (kN/m}^2)$$

List of symbols

A	area	m^2
C	concentration	kmol/m^3

C^*	concentration in equilibrium with P_a in eqn. 5-20	kmol/m^3
ΔC	concentration difference	kmol/m^3
d	wetted wall diameter	mm, m
D, D'	diffusivity	m^2/s
F	function in eqn. 5-13 and Fig. 5-5	K/N
G_m	molar gas rate	kmol/s
H	Henry's Law constant	(kN/m^2)/(kmol/m^3)
k_G	gas phase transfer coefficient	kmol/m^3 s (kN/m^2)
k_L	liquid phase transfer coefficient	m/s
K	constant	
K_G	overall mass transfer coefficient	kmol/m^2 s(kN/m^2)
K_L	overall mass transfer coefficient	m/s
L_m	molar liquid rate	kmol/s
M	molecular weight	kg/kmol
N	quantity transferred	kmol/m^2 s
P	pressure	kN/m^2
P_a	partial pressure	kN/m^2
P_{bm}	log mean partial pressure	kN/m^2
P_T	total pressure in eqn. 5-5	N/m^2
P^*	partial pressure in equilibrium with C_a in eqn. 5-19	kN/m^2
R	universal gas constant	kN/m/kmol K = ks/kmol K
Re	Reynolds Number	—
Sc	Schmidt Number	—
t	time	s
T	temperature	K
T_B	normal boiling temperature	K
v	specific volume	m^3/kg
V	solute molal volume	m^3/kmol
x	mol fraction in liquid phase	—
z	distance	mm, m
y	mole fraction in vapour phase	—
Z	distance	
α	constant in Example 5-9	—
ρ	density	kg/m^3
μ	viscosity	N s/m^2
ϕ	function in eqns. 5-6 to 5-8	—

Basic mass transfer

Subscripts
a, b   components a and b
1, 2   components 1 and 2
C   critical
G   gas
i   interface
L   liquid
t, b   toluene and benzene in Example 5-3

Further reading

COULSON, J. M., RICHARDSON, J. F., BACKHURST, J. R. &
  HARKER, J. H. *"Chemical Engineering"* Vol. I, 3rd Ed. Pergamon
  Press, Oxford (1978)
NORMAN, W. S. *"Absorption, Distillation and Cooling Towers"*
  Longmans, London (1961).
PERRY, J. H. Ed. *"Chemical Engineers Handbook"* 4th Ed.
  pp. 14–12 – 14–24 McGraw-Hill Book Co. New York (1963).
TREYBAL, R. E. *"Mass Transfer Operations"* McGraw-Hill Book Co.
  New York (1955).

Problem 5-1: In the experimental determination of the diffusivity of benzene using the system of Example 5-1, the following data were obtained:

   temperature = 295 K
   pressure = 100 kN/m^2
   density of benzene = 800 kg/m^3
   duration of experiment = 30 ks
   vapour pressure of benzene at 295 K = 13·3 kN/m^2
   initial height of benzene = 20 mm from the top of the tube
   final height of benzene = 25·8 mm.
Estimate the diffusivity of benzene.

$$(8·0 \times 10^{-6} \text{ m}^2/\text{s})$$

Problem 5-2: In an air/carbon dioxide mixture at 298 K and 202·6 kN/m^2, the concentrations of CO_2 at two planes 3·0 mm apart are 15% and 25%. The diffusivity of CO_2 at 101·3 kN/m^2 is 1·64 x 10^{-5} m^2/s. Calculate the rate of transfer of CO_2 across the two planes assuming:
   (a) equimolecular counter diffusion, and
   (b) diffusion of CO_2 through a stagnant air layer

$$(2·235 \times 10^{-5} ; 2·795 \times 10^{-5} \text{ kmol/s m}^2)$$

Problem 5-3:   Ammonia is diffusing through a stagnant layer of air 1 mm thick at a temperature of 293 K and at atmospheric pressure. At one boundary, the gas contains 50 vol. % of ammonia and at the other, the concentration of ammonia may be considered negligible. If the diffusivity of ammonia under these conditions is $1 \cdot 8 \times 10^{-5}$ m^2/s, calculate the rate of diffusion of ammonia through the layer.

$$(5 \cdot 2 \times 10^{-4} \text{ kmol/sm}^2)$$

Problem 5-4:   Hydrogen chloride is diffusing through an inert air film $1 \cdot 0$ mm thick at 293 K and 100 kN/m^2. Estimate the effect of increasing the pressure to 1 MN/m^2 on the rate of diffusion if the concentration of hydrogen chloride at one boundary is: (a) 20 kN/m^2 (b) 20% by volume.

$$(N_{a2} = 9 \cdot 27\% N_{a1}; \text{ no change}).$$

Problem 5-5:   A gas is being transferred across a stagnant air film at a total pressure of 100 kN/m^2. The partial pressure of the gas is 40 kN/m^2 at one boundary of the film and 10 kN/m^2 at the other. If the partial pressures remain constant, calculate the total pressure to double the transfer rate of the gas.

$$(63 \cdot 8 \text{ kN/m}^2)$$

6. Heat Transfer Equipment

6.1 Introduction

Some of the most important engineering problems involving heat transfer are concerned with the design of heat exchangers, that is, devices that continuously transfer heat between two fluid streams. In the majority of practical applications, the fluids are separated by an intervening wall which is the actual heat transfer surface of the exchanger. This type of unit is called a recuperative exchanger.

The other class of heat exchanger which is of industrial importance is the regenerator. Their operation is characterized by the internal surfaces, which may be variously metal, brick or ceramic, etc., being alternately exposed to the two fluids between which heat is to be transferred. Thus the material of the exchanger alternately stores heat extracted from the hot fluid and then delivers it to the cooler fluid. In the majority of applications, the operation of regenerators is naturally intermittent; however, continuous regenerators are available for certain situations.

6.2 Simple forms of heat exchanger

One of the simplest forms of recuperative exchanger is the so-called double-pipe exchanger. This arrangement is shown schematically in Fig. 6-1. One of the fluids passes through the tube whilst the second fluid flows through the annular space. If the fluids flow through the

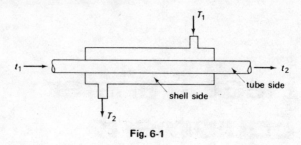

Fig. 6-1

exchanger in the same direction, then the unit is said to operate in parallel or co-current flow. Flow in opposite directions is called counter-current flow. The duty Q of an exchanger can be calculated from the change/unit time in sensible heat content of a fluid being heated or cooled, and/or the latent heat extracted or supplied in condensation or evaporation processes respectively. In terms of heat transfer, the duty of the exchanger can be described by the simple equation

$$Q = UA\theta_m \tag{6-1}$$

In this equation, U is the overall heat transfer coefficient referred to area A, which is the heat transfer surface required in the exchanger, and θ_m is the appropriate mean temperature difference between the two fluids. Since we are usually concerned with overall heat transfer coefficients across a tube wall, the overall coefficient U can be referred to either the inside or outside surface of the tube.

6.3 Mean temperature difference in simple heat exchange configurations

The mean temperature difference for a number of simple heat exchange configurations will now be considered. These are shown in Figs. 6-2 (a) to (e). For the parallel flow arrangement (Fig. 6-2(d)), consider a heat balance over the complete heat exchanger, and also over an element of the exchanger. (As an additional exercise the reader should note down the assumptions which are implicit in the following analysis):

$$Q = (mc)_T(T_1 - T_2) \tag{6-2}$$

$$Q = (mc)_t(t_2 - t_1) \tag{6-3}$$

$$Q = UA\,\theta_m \tag{6-4}$$

144

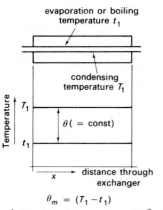

a) *Simple condenser/evaporator* *

evaporation or boiling temperature t_1

condensing temperature T_1

$\theta (= const)$

$\theta_m = (T_1 - t_1)$

distance through exchanger

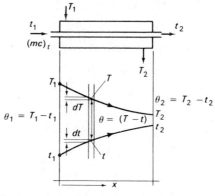

mass flowrate × specific heat $= (mc)_T$

$\theta_1 = T_1 - t_1$

$\theta = (T - t)$

$\theta_2 = T_2 - t_2$

d) *Parallel flow exchanger*

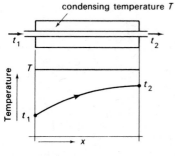

condensing temperature T

b) *Condenser/heater*

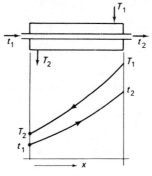

e) *Counter-current flow exchanger*

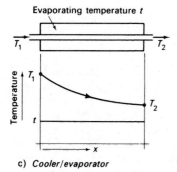

Evaporating temperature t

c) *Cooler/evaporator*

*(pure fluids) no subcooling of condensate, no superheating of vapour

Fig. 6-2

and

$$dQ = - (mc)_T \, dT \tag{6-5}$$

$$dQ = (mc)_t \, dt \tag{6-6}$$

$$dQ = U \, dA(T - t) = U \, dA\theta \tag{6-7}$$

From eqns. 6-5 and 6-6

$$dQ \left(- \frac{1}{(mc)_T} - \frac{1}{(mc)_t} \right) = dT - dt = d(T - t) = d\theta$$

From eqn. 6-7

$$U\theta dA \left(- \frac{1}{(mc)_T} - \frac{1}{(mc)_t} \right) = d\theta$$

$$U \left(- \frac{1}{(mc)_T} - \frac{1}{(mc)_t} \right) dA = \frac{d\theta}{\theta}$$

Integration gives

$$U \left(- \frac{1}{(mc)_T} - \frac{1}{(mc)_t} \right) A = \ln(\theta_2/\theta_1) \tag{6-8}$$

where $\theta_2 = T_2 - t_2, \theta_1 = T_1 - t_1$

From eqns. 6-2 and 6-3

$$Q \left(- \frac{1}{(mc)_T} - \frac{1}{(mc)_t} \right) = -(T_1 - T_2) - (t_2 - t_1)$$

$$= (T_2 - t_2) - (T_1 - t_1) = \theta_2 - \theta_1 \tag{6-9}$$

From eqns. 6-8 and 6-9

$$\theta_m = (\theta_2 - \theta_1)/\ln(\theta_2/\theta_1) \tag{6-10}$$

This is the logarithmic mean temperature difference (*LMTD*). It is left to the reader to show that the same expression applies for the mean temperature difference for the counter-current flow configuration, as well as the cooler/evaporator and condenser/heater shown in Fig. 6-2.

Equations 6-2, 6-3, 6-4 and 6-10 can be used either to predict the performance of a given exchanger, or to size an exchanger for a given duty. However, unless the four terminal temperatures of the exchange are known, the use of the logarithmic mean temperature difference will involve a trial and error solution.

A more convenient procedure results if a number of dimensionless parameters is defined as follows.

Heat exchanger effectiveness ε

This is the ratio of the actual rate of heat transfer Q to the maximum rate of heat transfer permitted by the Second Law of Thermodynamics. This maximum rate is given by

$$Q_{max} = (mc)_{min}(T_1 - t_1)$$

where $(mc)_{min}$ is the smaller of the two heat capacity rates, and T_1 and t_1 are the inlet temperatures of the hot and cold fluids, respectively. Then

$$\epsilon = \frac{(mc)_T(T_1 - T_2)}{(mc)_{min}(T_1 - t_1)} = \frac{(mc)_t(t_2 - t_1)}{(mc)_{min}(T_1 - t_1)} \tag{6-11}$$

The number of transfer units, NTU.

By definition

$$NTU = \frac{UA}{(mc)_{min}} \tag{6-12}$$

The NTU is the measure of the size of a heat exchanger from the point of view of heat transfer.

(Ratio of heat capacity rates (capacity ratio), $(mc)_{min}/(mc)_{max}$)

For a particular flow arrangement, the effectiveness is a function of the number of transfer units and the capacity ratio only.

Consider the case of the parallel flow from eqn. 6-8, with $(mc)_T = (mc)_{min}$

$$(T_2 - t_2)/(T_1 - t_1) = \exp(-NTU(1 + (mc)_{min}/(mc)_{max}))$$

also from eqn. 6-11

$$(1/(mc)_{min} + 1/(mc)_{max})\epsilon = (T_1 - T_2 + t_2 - t_1)/(T_1 - t_1)/(mc)_{min}$$

hence

$$(1 + (mc)_{min}/(mc)_{max})\epsilon = 1 - \exp(-NTU(1 + (mc)_{min}/(mc)_{max}))$$

from which

$$\epsilon = \frac{1 - \exp(-NTU(1 + (mc)_{min}/(mc)_{max}))}{(1 + (mc)_{min}/(mc)_{max})} \tag{6-13}$$

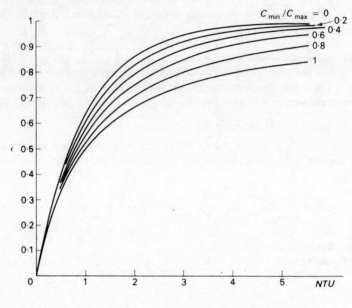

Fig. 6-3

It can be shown that for a counter-current exchanger the appropriate expression for ϵ is

$$\epsilon = \frac{1 - \exp\left(-NTU\left(1 - (mc)_{min}/(mc)_{max}\right)\right)}{1 - \left((mc)_{min}/(mc)_{max}\right)\exp\left(-NTU\left(1 - (mc)_{min}/(mc)_{max}\right)\right)}$$

(6-14)

For convenience, eqn. 6-14 and the appropriate equation for an exchanger having one shell pass with two tube passes are shown in Figs. 6-3 and 6-4 respectively.

Consider the cases when $(mc)_{min}/(mc)_{max}$ are equal to 0 and 1 respectively. When this ratio is zero, one of the fluids passing through the exchanger is undergoing a change of phase (i.e. a condenser or evaporator). In this case, eqns. 6-13 and 6-14 both reduce to

$$\epsilon = 1 - \exp\left(-NTU\right)$$

When $(mc)_T = (mc)_t$, ie a capacity ratio of 1, for parallel or co-current flow

$$\epsilon = \frac{1 - \exp\left(-2NTU\right)}{2}$$

148

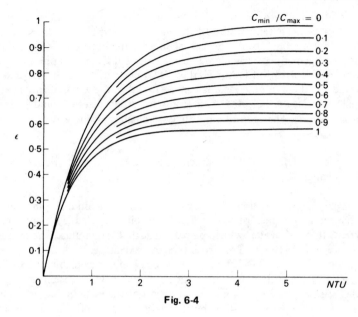

Fig. 6-4

and for counter-current flow

$$\epsilon = \frac{NTU}{1 + NTU}$$

It should be noted that as $NTU \to \infty$, $\epsilon \to 1$ for true counter flow operation, whilst the limit is $0 \cdot 5$ for parallel flow configuration.

6.4 More complex heat exchanger flow configurations

In the majority of heat exchanger designs, the magnitude of the required duty would mean the use of double-pipe heat exchangers of impractical length. The answer to this problem is usually the use of the conventional shell and tube exchanger. One fluid flows through the tubes, whilst the other fluid is passed through the shell and flows over the outside surface of the tubes. In order to ensure that the shell side fluid will flow across the tubes and induce satisfactory heat transfer rates, the shell is equipped with baffles. By suitable header design, the tube side fluid can be made to make one, two or more passes through the shell.

For a single tube pass single shell pass exchanger the appropriate mean

149

temperature difference for countercurrent or parallel flow is the logarithmic mean temperature difference discussed above. When using other flow configurations the actual mean temperature difference for that arrangement is calculated using the *LMTD* and applying an appropriate correction factor F (the *LMTD* correction factor). Detailed analyses of these configurations can be found in the work of Bowman, Muelle and Nagle. Selected data from this study can be found in the further reading. The duty of a multipass exchanger is, therefore

$$Q = UA\theta_m F \tag{6-15}$$

The equivalent ϵ, NTU and $(mc)_{min}/(mc)_{max}$ characteristics for a 1/2 exchanger are shown in Fig. 6-4.

Example 6-1: Oil flows through the tube of a double pipe heat exchanger at the rate of $0·189$ kg/s. The oil is cooled by a counter-current flow of water which passes through the annulus. The water flowrate is $0·151$ kg/s. The oil enters the exchanger at 422 K and is required to leave at 344 K when the cooling water is available at 283 K. If the tube has a mean diameter of $12·7$ mm and its wall presents a negligible resistance to heat transfer what length of tube will be required for this duty? Compare this with the length required if both liquids passed through the exchanger in the same direction.

Oil side heat transfer coefficient	2270 W/m² K
Water side heat transfer coefficient	5670 W/m² K
Specific heat of oil	$2·18$ kJ/kg K
Specific heat of water	$4·19$ kJ/kg K

Solution

Inlet oil temperature   422 K   exit temperature   344 K
Inlet water temperature   283 K   exit temperature t_2 K

A heat balance over the exchanger gives

duty = mass flowrate x specific heat x temperature change
$$0·189 \times 2·18 \times 10^3(422 - 344) = 0·151 \times 4·19 \times 10^3(t_2 - 283)$$

therefore

$$t_2 = \frac{0·189 \times 2·18 \times 10^3 (422 - 344)}{0·151 \times 4·19 \times 10^3} + 283 = 333·8 \text{ K}$$

For counter-current operation:

Oil/water inlet temperature difference $\theta_1 = (344 - 283) = 61$ K
Oil/water exit temperature difference $\theta_2 = (422 - 333·8) = 88·2$ K

and

$$\theta_m = \frac{\theta_2 - \theta_1}{\ln(\theta_2/\theta_1)} \text{ or } \frac{\theta_1 - \theta_2}{\ln(\theta_1/\theta_2)}$$

It is usually more convenient to use the expression such that the ratio of the temperature differences is greater than unity; substituting

$$\theta_m = (88{\cdot}2 - 61)/\ln(88{\cdot}2/61) = 73{\cdot}9 \text{ K}$$

Let U be the overall heat transfer coefficient, then

$$1/U = 1/2270 + 1/5670$$

hence

$$U = 1622 \text{ W/m}^2 \text{ K}$$

The exchanger duty

$$Q = 0{\cdot}189 \times 2{\cdot}18 \times 10^3 (422 - 344) = UA\theta_m = 1622 \times 73{\cdot}9 \times A$$

therefore

$$A = 0{\cdot}189 \times 2{\cdot}18 \times 10^3 (422 - 344)/(1622 \times 73{\cdot}9) = 0{\cdot}268 \text{ m}^2$$

However

$$A = \pi DL$$

then

$$L = 0{\cdot}268/(\pi \times 12{\cdot}7 \times 10^{-3}) = 6{\cdot}71 \text{ m}$$

For parallel flow:

$$\theta_1 = (422 - 283) = 139 \text{ K}$$
$$\theta_2 = (344 - 333{\cdot}8) = 10{\cdot}2 \text{ K}$$
$$\theta_m = (139 - 10{\cdot}2)/\ln(139/10{\cdot}2) = 49{\cdot}3 \text{ K}$$

Since the area required is inversely proportional to θ_m, then in the present problem the length of tube required is inversely proportional to θ_m

Length required for parallel flow arrangement is

$6{\cdot}71 \times 73{\cdot}9/49{\cdot}3 = 10{\cdot}05$ m (that is, approximately 50% more tube is required)

It should be noted that we have assumed that the modified temperature distribution throughout the exchanger does not affect the value of the heat transfer coefficient.

Example 6-2: A condenser is to be designed to deal with 7·5 kg/s of steam. The steam will have a dryness fraction of 0·9 and will be condensed at a pressure of 4·13 kN/m². Cooling water is available at 286 K; however, for economic reasons the temperature rise through the condenser must be limited to 10 K. Previous experience indicates that a water velocity of 1·5 m/s should be maintained in the tubes. If the exchanger is to have two tube passes calculate the number of tubes required for this duty, and their length.

Additional data:

Tubes 20 mm O.D. x 1·5 mm wall

Overall heat transfer coefficient based on the external area of the tube = 3520 W/m²K

Solution: The dryness fraction α of a vapour is defined by

$$\alpha = (H - H_f)/H_{fg} \tag{6-16}$$

where H is the specific enthalpy of the wet vapour H_f is the specific enthalpy of the saturated liquid, and H_{fg} is the latent heat of evaporation. Therefore α represents the fraction of the latent heat possessed by the vapour. For a dry saturated vapour the enthalpy is H_g, $H_g - H_f = H_{fg}$ and $\alpha = 1$.

From steam tables:—

$$H_{fg} = 2432 \text{ kJ/kg}$$
$$H_f = 123\cdot7 \text{ kJ/kg}$$

from eqn. 6-16

$$(H - H_f) = \alpha H_{fg}$$

In order to reduce 1 kg of wet vapour to 1 kg saturated liquid the exchanger must extract αH_{fg}.

Exchanger duty is

$$7\cdot56 \times 0\cdot9 \times 2432 \times 10^3 = 16\cdot42 \times 10^6 \text{ J/s} (\equiv W)$$

If the mass flowrate of water through the exchanger is M kg/s, then a heat balance gives

$$M \times 4\cdot186 \times 10^3 \times 10 = 16\cdot42 \times 10^6$$

from which

$$M = 392 \text{ kg/s}$$

The density of the cooling water is 996·5 kg/m³, therefore

volumetric water flowrate = $(392/996 \cdot 4) = 0 \cdot 393$ m^3/s

Let the total number of tubes in the exchanger be N then the number of tubes per pass is $N/2$.

Cross sectional area of tube/pass

$$a = \pi(17 \times 10^{-3})^2/4 = 2 \cdot 27 \times 10^{-4} \, m^2$$

Flow area per pass is

$$a \, N/2$$

Volumetric flowrate is

$1 \cdot 5 \, a \, N/2$ (velocity × flow area per pass) = $0 \cdot 393$ m^3/s

then

$$N = 0 \cdot 393 \times 2/(1 \cdot 5 \times 2 \cdot 27 \times 10^{-4}) = 2308 \text{ tubes}$$

Steam/cooling water temperature difference at inlet

$$\theta_1 = (302 \cdot 5 - 286) = 16 \cdot 5 \text{ K}$$

Steam/cooling water temperature difference at outlet

$$\theta_2 = (302 \cdot 5 - 296) = 6 \cdot 5 \text{ K}$$

then

$$\theta_m = \theta_1 - \theta_2/\ln(\theta_1/\theta_2) = 10/\ln 2 \cdot 54 = 10 \cdot 7 \text{ K}$$

but

$$Q = UA \, \theta_m$$

from which

$$A = Q/(U\theta_m) = 16 \cdot 42 \times 10^6/(3520 \times 10 \cdot 7) = 435 \text{ m}^2$$

Outside surface area of tubes/m length

$$a_s = \pi \times 20 \times 10^{-3} = 6 \cdot 29 \times 10^{-2} \text{ m}^2/m$$

Let the length of each tube be L(m) then

$$A = 435 = Na_sL$$

and

$$L = 2 \cdot 91 \text{ m}$$

Example 6-3: A process liquor at $299 \cdot 5$ K is to be heated using water at 366 K available from another part of the plant. A heat exchanger having an installed area of $8 \cdot 1 \text{ m}^2$ is to be used for this operation, with the liquids flowing in true counter-current manner. The flowrates of the process liquor and water will be $3 \cdot 1$ and $1 \cdot 1$ kg/s, respectively. Previous experience indicates that an overall heat transfer coefficient of $454 \text{ W/m}^2 \text{ K}$ will apply. The specific heat of the process liquor is $2 \cdot 1$ kJ/kg K, that of water $4 \cdot 2$ kJ/kg K. Estimate the exit temperatures of its liquor and water streams.

Solution: Thermal capacity of process liquor stream is

mass flowrate x specific heat $= 3 \cdot 1 \times 2 \cdot 1 \times 10^3 = 6 \cdot 51 \times 10^3 \text{ J/s K} = (mc)_t$

Thermal capacity of water stream is

$$1 \cdot 1 \times 4 \cdot 2 \times 10^3 = 4 \cdot 62 \times 10^3 \text{ WK} = (mc)_T$$

Capacity ratio is

$$(mc)_{min}/(mc)_{max} = (mc)_T/(mc)_t = 4 \cdot 62/6 \cdot 51 = 0 \cdot 71$$

Number of transfer units is

$$AU/(mc)_{min} = 8 \cdot 1 \times 454/4 \cdot 62 \times 10^3 = 0 \cdot 8$$

From Fig. 6-3

$$\epsilon = 0 \cdot 47$$

but

$$\epsilon = (mc)_t(t_2 - 299 \cdot 5)/(mc)_{min}(T_1 - t_1)$$
$$= 6 \cdot 51 \times 10^3(t_2 - 299 \cdot 5)/4 \cdot 62 \times 10^3(366 - 299 \cdot 5) = 0 \cdot 47$$

Therefore, exit liquor temperature t_2 is $321 \cdot 7$ K
Also

$$\epsilon = (mc)_T(366 - T_2)/(mc)(T_1 - t_1)$$
$$= 4 \cdot 62 \times 10^3(366 - T_2)/4 \cdot 62 \times 10^3(366 - 299 \cdot 5) = 0 \cdot 47$$

Therefore, exit water temperature is $334 \cdot 7$ K.

Example 6-4:   It is proposed to condense low pressure steam in an existing heat exchanger which has an installed area of 20 m^2. This operation is to be carried out whilst conveniently heating the feedstock required by another process. This feedstock will be available at 313 K, and the flowrate will be $0 \cdot 917$ kg/s. Estimate the quantity of steam condensed and the exit temperature of the feedstock if an overall heat transfer coefficient of 125 $W/m^2 K$ can be achieved. If this coefficient could be doubled by modification of feedstock flow through the exchanger how would this affect the performance of the condenser? The steam will be condensed at atmosphere pressure (T_s = 373 K).

Solution:
    Let the exit temperature of the feedstock be t K, then the duty of the exchanger is

$$0{\cdot}917 \times 4000\,(t - 313) = 3666\,(t - 313)\,W = UA\theta_m$$

In this case, the mean temperature difference θ_m is the *LMTD*, then

$$\theta_m = \frac{(373 - 313) - (373 - t)}{\ln(373 - 313)/(373 - t)} = (t - 313)/\ln[60/(373 - t)]$$

substituting

$$3666(t - 313) = 125 \times 20 \times (t - 313)/\ln[60/(373 - t)]$$

then

$$\ln[60/(373 - t)] = 2500/3666 = 0{\cdot}68$$

from which

$$t = 342{\cdot}6 \text{ K}$$

However, the duty of the exchanger is

$$3666\,(t - 313) = 3666 \times 29{\cdot}6\,W = \text{mass of steam condensed/}$$
$$\text{second} \times \text{latent heat.}$$

This assumes that the steam was initially just saturated and that the condensate leaves the exchanger without subcooling.
Latent heat of steam at atmospheric pressure is 2257 kJ/kg
Then, rate of condensation is

$$\frac{3666 \times 29{\cdot}6}{225{\cdot}7 \times 10^4} = 0{\cdot}048 \text{ kg/s}$$

If the overall heat transfer coefficient is doubled, i.e. U becomes 250, let the feedstock exit temperature be t'. A similar calculation gives

$$\ln[60/(373 - t')] = 5000/3666 = 1{\cdot}364$$

from which

$$t' = 357 \cdot 6 \text{ K}$$

Then, rate of condensation is

$$\frac{3666(357 \cdot 6 - 313)}{225 \cdot 7 \times 10^4} = 0 \cdot 072 \text{ kg/s}$$

i.e. doubling the overall coefficient produces approximately 50% in exchange duty. The reader might care to consider the effect of doubling the coefficient yet again, i.e. to 500 W/m^2 K.

Example 6-5: A gas is to be heated from temperature T_1 to T_2 K by passing it through a steam heated tube. If the mean temperature difference and heat transfer coefficient between the tube wall and the gas is θ_m and h respectively show that

$$St = \left(\frac{h}{\rho V c_p}\right) \left(\frac{T_2 - T_1}{\theta_m}\right) \left(\frac{D}{4L}\right)$$

where D and L are inside diameter and length of the tube respectively. V the mean flow velocity of the gas, and ρ and c_p the mean density and specific heat of the gas respectively.

A preheater is to be designed to heat $1 \cdot 8$ kg/s of air from 288 to 348 K using steam condensing at 373 K as the heating medium. The maximum permitted pressure drop across the tubes of the exchanger, which are 25 mm inside diameter, is limited to $7 \cdot 5 \times 10^2$ N/m^2. Calculate the number of tubes required and their length.

Additional data:
  Density of air at 318 K = $1 \cdot 11$ kg/m^3
  Viscosity of air at 318 K = $0 \cdot 192 \times 10^{-4}$ Ns/m^2

  Stanton number $St = 0 \cdot 046 \, Re^{-\frac{1}{4}}$
  Friction factor $f = 0 \cdot 079 \, Re^{-\frac{1}{4}}$

Solution: (Heat gained by gas in passing through the tube) = (mass flowrate x specific heat x rise in temperature)

$$= \frac{\pi D^2}{4} V \rho c_p (T_2 - T_1)$$

(Heat transferred from the tube surface to the gas) = (surface area x heat transfer coefficient x mean temperature difference) = $\pi DL \times h \times \theta_m$.

Therefore

$$\frac{\pi D^2}{4} \, V\rho c_p(T_2 - T_1) = DL \, h \cdot \theta_m$$

A rearrangement gives the required expression

$$\left(\frac{h}{\rho V c_p}\right) = St = \left(\frac{T_2 - T_1}{\theta_m}\right)\left(\frac{D}{4L}\right)$$

The reader should note, and check, that the terms in this equation are conveniently dimensionless. The equation also suggests a simple method of measuring either overall or limiting film coefficient in heat exchangers. Since

$$\Delta P = 2f \frac{L}{D} \, \rho V^2 \quad \text{(see chapter 3)}$$

then

$$\frac{D}{L} = \frac{2f\rho V^2}{\Delta P}$$

and

$$St = \left(\frac{T_2 - T_1}{\theta_m}\right)\left(\frac{f\rho V^2}{2\Delta P}\right)$$

or

$$V = \left[\left(\frac{St}{f/2}\right)\frac{\theta_m}{(T_2 - T_1)} \left(\frac{\Delta P}{\rho}\right)\right]^{\frac{1}{2}}$$

If $\Delta P = \Delta P_{max}$ and $V = V_{max}$

$$\theta_m = \frac{(373 - 288) - (373 - 348)}{\ln(373 - 288)/(373 - 348)} = 49 \text{ K}$$

and

$$V_{max} = \left(\frac{0 \cdot 046}{0 \cdot 0395} \times \frac{49}{60} \times \frac{7 \cdot 5 \times 10^2}{1 \cdot 11}\right)^{\frac{1}{2}}$$

$$= 25 \cdot 47 \text{ m/s}$$

157

Mass flowrate through a single tube is

$$\frac{\pi D^2}{4} \rho V_{max} = \frac{\pi (25)^2}{4} \; 10^{-6} \times 25.35 \times 1.11 = 138.1 \times 10^{-4} \text{ kg/s}$$

If the number of tubes in the exchanger is N, then

$$138.8 \times 10^{-4} \times N = 1.8 \text{ kg/s}$$

hence

$$N = 1.8/138.1 \times 10^{-4} = 130.3 \text{ say 131 tubes}$$

also

$$Re = \frac{D V_{max} \rho}{\mu} = \frac{2.5 \times 10^{-2} \times 25.47 \times 1.11}{0.192 \times 10^{-4}} = 3.68 \times 10^4$$

Then friction factor

$$f = 0.079 \, Re^{-\frac{1}{4}} = 0.0057$$

Since $\qquad \Delta P = 2f \dfrac{L}{D} \rho V_{max}^2$

$$\frac{L}{D} = \frac{\Delta P}{2f\rho V_{max}^2} = \frac{7.5 \times 10^2}{2 \times 0.0057 \times 1.11 \times (25.35)^2} = 92.2$$

then

$$L = 91.44 \times 2.5 \times 10^{-2} = 2.28 \text{ m}$$

Example 6-6:   A food processing plant requires 4.17 kg/s of purified water at 343 K. This water is available from a purification process at 323 K and it is to be heated in a conventional shell and tube exchanger using 8.34 kg/s of untreated process water. This untreated water can be made available at four different temperatures, 383, 373, 363 and 353 K by different process flow arrangements; however, its cost increases with temperature. Calculate the surface areas required for exchangers having one shell pass, two tube passes operating with the given inlet temperatures. The overall heat transfer coefficient for the exchangers may be taken as 1500 W/m^{2}K. The specific heat of the purified and untreated water is 4.18 kJ/kg K.

Solution   Thermal capacity of the process water stream (C) is

mass flowrate x specific heat = $8.34 \times 4.18 \times 10^3$ W/K.

Thermal capacity of purified water stream is

$4 \cdot 17 \times 4 \cdot 18 \times 10^3$ W/K.

Capacity ratio is

$$C_{min}/C_{max} = 4 \cdot 17 \times 4 \cdot 18 \times 10^3/8 \cdot 34 \times 4 \cdot 18 \times 10^3 = 0 \cdot 5$$

Heat exchanger effectiveness

$$\epsilon = \frac{\text{exchanger duty}}{C_{min}(T_1 - t_1)} = \frac{4 \cdot 17 \times 4 \cdot 18 \times 10^3 (t_2 - t_1)}{4 \cdot 17 \times 4 \cdot 18 \times 10^3 (T_1 - t_1)}$$

$$= (343 - 323)/(T_1 - 323) = 20/(T_1 - 323)$$

For particular value of the process water inlet temperature T_1, can be calculated. Using this value of ϵ and the capacity ratio, the appropriate value of NTU can be determined from Fig. 6-4. Since $NTU = A \cdot U/C_{min}$ the area required A can be calculated. These calculations have been performed for the specified values of inlet temperature T_1 and the results are shown in Fig. 6-5.

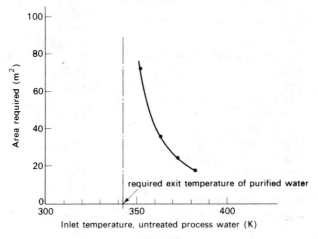

Fig. 6-5

T_1 (K)	C_{min}/C_{max}	ϵ	NTU	A (m^2)
383	0·5	0·33	0·45	18·8
373	0·5	0·4	0·6	25·1
363	0·5	0·5	0·875	36·6
353	0·5	0·66	1·75	73·5

The area required, as a function of inlet temperature T_1, is shown in Fig. 6-5. The graph shows the rapid increase in area required as the inlet temperature T_1 approaches the outlet temperature of the purified water. Since the cost of providing the heating water increases with temperature, whilst the surface area required in the appropriate exchanger decreases, the designer is faced with the necessity of determining the optimum solution to this problem. That is to say, he must balance capital and running costs such that the total costs are minimized. In practice, optimization is not quite so simple since many other factors influence the optimum solution, though these results indicate the essential elements of the problem.

List of symbols

a	cross sectional flow area of a tube	(m^2)
a_s	external surface area of tube/unit length	(m^2/m)
A	area	(m^2)
c	specific heat	$(J/kg\ K)$
C	thermal capacity of exchanger stream (mc)	(W/K)
D	diameter	(m)
F	*LMTD* correction factor	$(-)$
h	heat transfer coefficient	$(W/m^2 K)$
H	specific enthalpy	(J/kg)
k	thermal conductivity	$(W/m\ K)$
L	length of exchanger tube	(m)
M, m	mass flowrate	(kg/s)
N	number of tubes in exchanger	$(-)$
NTU	number of transfer units	$(-)$
p	pressure, Δp pressure drop	(N/m^2)
Q	exchanger duty	(W)
T, t	temperature	(K)
U	overall heat transfer coefficient	$(W/m^2 K)$
V	velocity	(m/s)
ϵ	heat exchanger effectiveness	$(-)$
θ_m	logarithmic mean temperature difference	(K)
μ	viscosity	(Ns/m^2)
ρ	density	(kg/m^3)

dimensionless numbers
f friction factor (see chapter 3)
Nu Nusselt number (hD/k)
Pr Prandtl number $(c\mu/k)$
Re Reynolds number $(D\,V\,\rho/\mu)$
St Stanton number $(h/c\,V\,\rho) = (Nu/Re\,Pr)$

Further Reading

FRAAS, H. P. and OZISIK, M. N. *"Heat Exchanger Design"* Wiley, New York, (1965).
KERN, D. Q. *"Process Heat Transfer"* McGraw-Hill, New York (1950).
KAYS, W. M. and LONDON, A. L. *"Compact Heat Exchangers"* McGraw-Hill, New York (1958).
McADAMS, W. H. "Heat Transmission" 3rd Edition McGraw-Hill, New York (1954).
JAKOB, M. *"Heat Transfer"* Vol. II Wiley, New York (1957).
MAYHEW, Y. R. and ROGERS, G. F. C. *"Thermodynamic and Physical Properties of Fluids"* S.I. Units Oxford University Press (1971).
HAYWOOD, R. W. *"Thermodynamic Tables in S.I. Units"* Cambridge University Press (1968).
SCHACK, A. *"Industrial Heat Transfer"* Chapman and Hall, London (1965).
HOLLAND, F. A., MOORES, R. M., WATSON, F. A., and WILKINSON, J. K. *"Heat Transfer"* Heinemann, London (1970)

Problem 6-1: A process water heater is to be designed using data from a heat exchanger already in service. This exchanger, which has one shell pass and two tube passes, has an effective area of 185 m², and is used to heat 13·9 kg/s of water. The temperature of the water is to be raised from 288 to 363 K using exhaust gases from a reactor. These gases enter the reactor at 613 K and leave at 423 K. Estimate the overall heat transfer coefficient under these conditions.

Additional data:— Specific heat of reactor gases 1·1 kJ/kg K
Specific heat of process water 4·2 kJ/kg K
(136·5 W/m²K)

Problem 6-2: If the exchanger required in Problem 6-1 is to operate in true counterflow, estimate the effective heat transfer area which should be specified. Water and gas flowrates, terminal temperatures and the anticipated overall heat transfer coefficient are to be taken as before.

$$(172 \cdot 7 \text{ m}^2)$$

Problem 6-3: A sea water/fresh water exchanger is to be constructed using tubes with an inside diameter of 20 mm. The sea water which will flow through the tubes, will enter at 305 K and leave at 328 K, the flowrate being $4 \cdot 17$ kg/s. The fresh water flowrate in the shell is to be $2 \cdot 78$ kg/s entering at 363 K. If the tube side velocity is to be $0 \cdot 35$ m/s calculate the number of tubes/pass and their length. The exchanger should have one shell pass, two tube passes. The density and specific heat of both sea water and fresh water may be taken as 960 kg/m^3, and $4 \cdot 2$ kJ/kg K respectively. The overall heat transfer coefficient based on the inside area of the tubes is estimated at 2100 W/m^{2}K.

$$(40, 1 \cdot 6 \text{ m})$$

Problem 6-4: An existing heat exchanger having 100 tubes, $12 \cdot 5$ mm bore, 15 mm outside diameter by $1 \cdot 5$ m long, is to be used to heat $1 \cdot 89$ kg/s water from 299 K to 322 K. The heating medium will be steam condensing at 373 K in the shell, the film coefficient for the condensate film being estimated at 8500 W/m^{2}K. The heat transfer coefficient in the tube is given by the equation

$$hD/k = 0 \cdot 023(D.V. \rho/\mu)^{0 \cdot 8}(C\mu/k)^{0 \cdot 3}$$

If the water makes a single pass through the exchanger, will the exchanger be suitable for this duty? The relevant properties of water at $310 \cdot 5$ K are

$k = 0 \cdot 62$ W/m K $\qquad \rho = 998$ kg/m^3

$C = 4 \cdot 2$ kJ/kg K $\qquad \mu = 6 \cdot 9 \times 10^{-4}$ Ns/m^2

$$(\text{yes})$$

Problem 6-5: Tubes 20 mm inside diameter are to be used to heat air from $288 \cdot 5$ K to $338 \cdot 5$ K. The tube wall temperature will be maintained constant at 383 K. If the pressure drop across the tubes is not to exceed 240 mm W. G., calculate the maximum tube velocity and the tube length. It may be assumed that $f = 0 \cdot 079 \, Re^{-\frac{1}{4}}$ and $St = 0 \cdot 046 \, Re^{-\frac{1}{4}}$.

Air properties: Density = $1 \cdot 1$ kg/m^3

Viscosity 19×10^{-6} Ns/m^2

Density of water = 998 kg/m^3 $\qquad\qquad (57 \cdot 4 \text{ m/s}, 1 \cdot 31 \text{ m})$

Problem 6-6: A heat exchanger is equipped with tubes 9·6 mm inside diameter and 1·5 m long. Air is to be passed through these tubes, entering at 293 K, and it will be heated by steam condensing at 373 K on the outside of the tubes. If the velocity of air through the tubes is to be 20 m/s, estimate the exit temperature of the air, and the pressure drop across the tubes. The density and viscosity of the air may be taken as 1·0 kg/m³ and 1·7 x 10⁻⁵ Ns/m² respectively. The air side coefficient may be considered limiting. The following information may be used as required.

$$f = 0·079 \, Re^{-\frac{1}{4}} \qquad \Delta P = 2f \frac{L}{D} \rho V^2$$

$$St = \frac{h}{\rho VC} = \frac{f}{2}$$

(365·6 K, 957·5 N/m²)

7. Distillation

7.1 Introduction

Distillation is the separation of the constituents of a liquid mixture by partial vaporization of the mixture and subsequent recovery of the vapour and the residue. The essential requirement for separation by distillation is that the composition of the vapour must be different from the composition of the liquid with which it is in equilibrium. Thus the vapour phase contains more than one component and the object of the distillation process is to recover one or more of these components. A distinction may be drawn between evaporation and distillation; in the former, usually only one component exists as vapour and this is frequently discarded leaving a more concentrated solution (or slurry) as product.

Two modes of operation are normally employed in distillation processes; in one, the liquid mixture is boiled and the vapours condensed without allowing any condensed liquid to return to the still, in the other, condensed vapours are returned as liquid reflux to form a continuous counter-current process. This latter process is known as fractionation or rectification and is of extreme importance. Both methods of operation are illustrated by means of examples in this chapter.

7.2 Vapour-liquid equilibria

7.2.1 Ideal systems

Vapour-liquid equilibrium data form the basis of all distillation operations and the most useful (though frequently not available) form of the data is the boiling point diagram. This is illustrated in **Fig. 7-1** and its application may be described as follows. The boiling points of

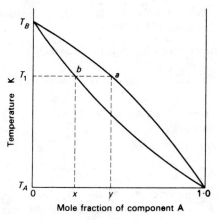

Mole fraction of component A

Fig. 7-1

pure A (the more volatile component) and pure B are represented by temperatures T_A and T_B respectively. Any point on the upper curve a has a vapour composition y and at the temperature T_1 it will just begin to condense to give a liquid of composition x. Thus, any two points (e.g. a and b) at the same temperature represent the compositions of liquid and vapour in equilibrium with each other. Any point above the top curve represents an all vapour mixture, all points below the bottom curve represent an all-liquid condition, and points between the two curves correspond to a liquid-vapour mixture.

If the boiling point diagram is unavailable for a particular system, it is normal to obtain equilibrium relationships from vapour pressure data using Raoult's Law and Dalton's Law, where ideal behaviour may be assumed. The method is illustrated in Example 7-1.

Example 7-1: The following vapour pressures were obtained for phenol and orthocresol.

Temperature (K)	Vapour pressure (kN/m^2)	
	o-cresol	phenol
387·0	7·70	10·0
387·9	7·94	10·4
388·7	8·21	10·8
389·6	8·50	11·2
390·3	8·76	11·6
391·1	9·06	12·0
391·9	9·40	12·4
392·7	9·73	12·9
393·3	10·00	13·3

165

Assuming Raoult's and Dalton's Laws apply, construct the following data for a total pressure of $10 \cdot 0 \ kN/m^2$.

(a) A temperature – composition diagram.
(b) An x-y diagram.
(c) Relative volatility against mole fraction of phenol in liquid.

Solution: If Raoult's Law and Dalton's Law apply, the following equations may be used

$$P = P_A + P_B \tag{7-1}$$

$$P_A = y_A \cdot P = x_A \cdot P_A^{\circ} \tag{7-2}$$

where P is the total pressure, P_A, P_B the partial pressures of components A and B, and P_A°, P_B° the vapour pressures of the pure components A and B.

Hence, from eqns. 7-1 and 7-2

$$x_A = (P - P_B^{\circ})/(P_A^{\circ} - P_B^{\circ}) \tag{7-3}$$

$$y_A = P_A^{\circ} \cdot x_A/P \tag{7-4}$$

Table 7-1

Temperature K	x_A (from eqn. 7-3)	y_A (from eqn. 7-4)	$\alpha = P_A^{\circ}/P_B^{\circ}$
387	1·0	1·0	1·300
387·9	0·837	0·871	1·310
388·7	0·691	0·746	1·315
389·6	0·556	0·622	1·318
390·3	0·437	0·506	1·324
391·1	0·320	0·384	1·325
391·9	0·200	0·248	1·319
392·7	0·085	0·110	1·326
393·3	0	0	1·330

The relative volatility is defined as P_A°/P_B° at a particular temperature and if by convention the subscript A refers to the more volatile component (in this case phenol), the above table may be completed and the required diagrams drawn.

A plot of the temperature – composition diagram is shown in

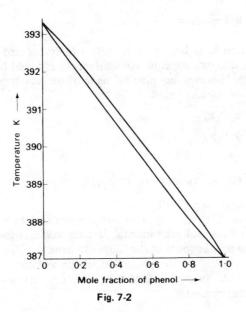

Fig. 7-2

Fig. 7-2 and the equilibrium diagram (x vs y) is shown in Fig. 7-3. The values obtained for the relative volatility vary between $1\cdot300$ and $1\cdot330$ and the average value of $1\cdot319$ could be used in subsequent calculations.

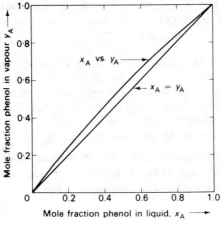

Fig. 7-3

167

7.2.2 Non-ideal systems

Deviation from ideal behaviour is fairly widespread and it should be emphasized that experimental equilibrium data should be used whenever possible. However, use may be made of the activity coefficient to correct for non-ideality as

$$Py_1 = \gamma_1 P_1 x_1 \tag{7-5}$$

$$Py_2 = \gamma_2 P_2 x_2 \tag{7-6}$$

and

$$\alpha_{12} = (y_1/x_1)/(y_2 x_2) = (\gamma_1 P_1)/\gamma_2 P_2 \tag{7-7}$$

where γ is the activity coefficient.

Activity coefficients vary with composition and, to a lesser extent, with temperature, and experimental data are available for a number of systems. Use may be made of the Gibbs–Duhem, the Van Laar, and the Margules equations for the prediction of the variation of the activity coefficient with composition, and the reader is referred to a suitable text on thermodynamics.

7.3 Distillation processes

7.3.1 Equilibrium flash distillation

In equilibrium or flash distillation, a batch of liquid is heated and the vapour and liquid are maintained in intimate contact until the two phases are in equilibrium with each other. The vapour is then withdrawn and condensed as product.

Example 7-2: An aqueous solution at its boiling point containing 10 mol % of ammonia is to be distilled to produce a distillate containing 25 mol % of ammonia. At equilibrium, the mole fraction of ammonia in the vapour is $6\cdot3$ times that in the liquid and the feed flow rate is $0\cdot1$ kmol/s. Calculate the number of moles of distillate obtainable from an equilibrium flash distillation.

Solution: In the equilibrium still, the liquid and vapour are in equilibrium with each other, i.e. the liquid product composition x_w is in equilibrium with the product vapour $y = 0\cdot25$.

As $y = 6\cdot3\,x_w$

$$x_w = 0\cdot25/6\cdot3 = 0\cdot0397$$

An overall mass balance will give the required distillate rate D

$$0\cdot1(0\cdot1) = D(0\cdot25) + (0\cdot1 - D)(0\cdot0397)$$

then

$$D = 0\cdot00603/0\cdot2103 = 0\cdot0287 \text{ kmol/s}$$

The application of equilibrium distillation is commonly found in the pipe stills of the petroleum industry where multicomponent mixtures are heated under pressure and the vapour flashed off at near equilibrium conditions from the superheated liquid.

7.3.2 Differential distillation

Simple, or differential, distillation consists of boiling a batch of the liquid mixture fed to a still pot and continuously removing and condensing the vapour as it is formed, thus enriching the liquid remaining in the still.

Example 7-3: 100 kmol of a mixture of A and B is fed to a simple still. The feed contains 50 mol% of A and a product containing 5 mol% of A is required. Calculate the quantity of product obtained. The equilibrium data are presented in Fig. 7-4.

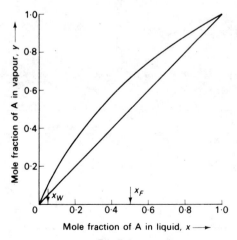

Fig. 7-4

Solution   If W_0 kmol of feed, composition x_F, are fed to the still initially, let W kmol of liquid, composition x, remain after a certain time. The vapour composition at this time is y and the total quantity of component A in the liquid is xW.

Now suppose a small amount of liquid dW is vaporized so that the liquid composition falls to $(x - dx)$ and the quantity falls to $(W - dW)$. The moles of A remaining in the still will be $(x - dx)(W - dW)$ and the quantity $y\,dW$ has been removed from the still.

A mass balance on component A gives

$$xW = (x - dx)(W - dW) + y\,dW \tag{7-8}$$

which, neglecting second order differentials gives

$$dW/W = dx/(y - x) \tag{7-9}$$

Integration of eqn. 7-9 between the limits of W_0 and W_w, the initial and final quantities in the still and between x_F and x_w, the initial and final compositions in the still, produces the Rayleigh Equation (eqn. 7-10)

$$\int_{W_s}^{W_0} dW/W = \ln(W_0/W_s) = \int_{x_s}^{x_F} dx/(y - x) \tag{7-10}$$

If we use the data from the equilibrium curve, this expression may be integrated graphically from a plot of $1/(y - x)$ against x as shown in the table below and Fig. 7-5

From Fig. 7-5

area under the curve = $3·725$

Hence

$\ln(100/W_s) = 3·725$ and $W_s = 2·4$ kmol

x	y	$(y - x)$	$1/(y - x)$
$0·50 = x_F$	$0·67$	$0·17$	$5·88$
$0·40$	$0·57$	$0·17$	$5·88$
$0·30$	$0·46$	$0·16$	$6·25$
$0·20$	$0·34$	$0·14$	$7·14$
$0·10$	$0·20$	$0·10$	$10·0$
$0·05 = x_w$	$0·10$	$0·05$	$20·0$

These data are presented in Fig. 7-5.

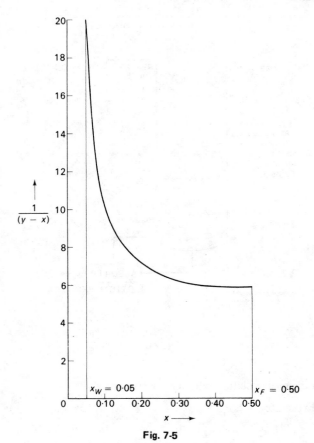

Fig. 7-5

7.3.3 Continuous distillation

(i) Mass and heat balances

A diagrammatic representation of a continuous rectification column is shown in Fig. 7-6. The vapour rising from the top plate is condensed and part returned to the column as reflux, the remainder forming the product D.

The mass balance over the whole column gives

$$F = D + W \tag{7-11}$$

171

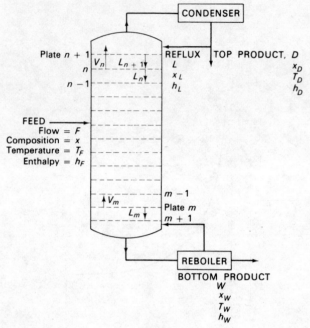

Fig. 7-6

and on the more volatile component (m.v.c.)

$$Fx_F = Dx_D + Wx_w \tag{7-12}$$

Hence

$$D = F(x_F - x_w)/(x_D - x_w) \tag{7-13}$$

and

$$W = F(x_D - x_F)/(x_D - x_w) \tag{7-14}$$

A mass balance above plate n in Fig. 7-6 gives

$$V_n = L_{n+1} + D \tag{7-15}$$

and for the m.v.c.

$$V_n y_n = L_{n+1} x_{n+1} + Dx_D \tag{7-16}$$

Hence

$$y_n = (L_{n+1}/V_n)x_{n+1} + (D/V_n)x_D \tag{7-17}$$

If the assumption of constant molal overflow is made, i.e. if the latent heat of vaporization of each component is approximately equal, we can write $L_{n+1} = L_n$ and obtain

$$y_n = (L_n/V_n)x_{n+1} + (D/V_n)x_D \qquad (7\text{-}18)$$

This is the equation of the operating time in the top (the rectifying) section of the column.

Similarly, for the bottom (stripping) section the operating line is

$$y_m = (L_m/V_m)x_{m+1} - (W/V_m)x_w \qquad (7\text{-}19)$$

As the reflux ratio R is defined as $R = L/D$, eqn. (7-18) may be written as

$$y_n = (R/(R+1))x_{n+1} + x_D/(R+1) \qquad (7\text{-}20)$$

Example 7-4:   If the system of Example 7-2 is now used in a continuous distillation column containing one ideal plate and a still, calculate the rate of product formation.

Solution   In the case of continuous distillation with one ideal plate and no reflux, the vapour above the plate has the composition $y = 0\cdot25$ and the liquid on the plate has $x_A = 0\cdot0397$. The system is shown in Fig. 7-7.

The lower operating time has the equation

$$y = (L_m/V_m)x_{n-1} - (W/V_n)x_w$$

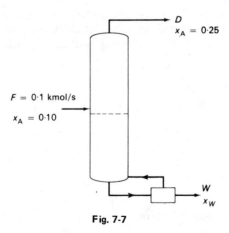

Fig. 7-7

In this case

$$L_m = F = 0·1 \text{ kmol/s and } V_m = V_n = D$$

Therefore,

$$y = (0·1/D) (0·0397) - (0·1 - D) (x_w/D)$$

where y represents the vapour rising from the still, and as

$$x_s = x_w = y_s/6·3$$
$$6·3\, x_w = y_s = 0·1(0·0397)/D - (0·1 - D)·x_w/D$$

from which

$$x_w = 0·003\ 97/(5·3D + 0·1)$$

An overall mass balance gives

$$(0·1 - D)x_w + 0·25D = 0·1(0·1)$$

that is

$$x_w = 0·01 - 0·25D/(0·1 - D)$$

Equating the two expressions for x_w provides an equation for D, thus

$$D^2 - 0·0241D - 0·000\ 455 = 0$$

Solving for D and rejecting the negative root gives the rate of product removal.

$$D = 0·0365 \text{ kmol/s.}$$

Heat Balances

If q_r is the heat added in the reboiler (kW), q_c the heat removed in the condenser (kW), H the enthalpy of vapour (kJ/kmol), and h the enthalpy of liquid (kJ/kmol) then, assuming no heat losses from the system

$$F·h_F + q_r = D·h_D + Wh_w + q_c \tag{7-21}$$

and over the top plate

$$V_t = L + D \tag{7-22}$$

$$V_t·y_t = L·x_t + D·x_D \tag{7-23}$$

$$V_t·H_t = q_c + L·h_L + D h_D \tag{7-24}$$

If a total condenser is used, $y_t = x_L = x_D$ and $h_L = h_D$.
From eqn. 7-24

$$q_c = V_t·H_t - (L + D)h_D \tag{7-25}$$

But

$$V_t = L + D \tag{7-22}$$

and

$$q_c = (L + D)H_t - (L + D)h_D \tag{7-26}$$

thus

$$\frac{q_c}{D} = \left(\frac{L}{D} + 1\right)(H_t - h_D) = (R + 1)(H_t - h_D) \tag{7-27}$$

q_r may now be calculated from eqn. 7-21.

Example 7-5: A continuous distillation column is to separate a mixture of A and B at atmospheric pressure. The feed rate to the column is $0 \cdot 1$ kmol/s, the feed is at its boiling point and contains 40 mol% of A. The top and bottom product specifications are 95 mol% and 5 mol% of A respectively. The reflux ratio is $3 \cdot 0$ kmol/kmol of top product and the reflux enters the column at 311 K. Given the following enthalpy data, estimate the condenser duty and the heat required in the reboiler.

$h_F = 10\ 300$ kJ/kmol

$h_D = 3340$ kJ/kmol

$H_t = 27\ 000$ kJ/kmol

$h_w = 14\ 700$ kJ/kmol

Solution
From eqn. 7-13

$$D = \frac{0 \cdot 1 (0 \cdot 4 - 0 \cdot 05)}{(0 \cdot 95 - 0 \cdot 05)} = 0 \cdot 0389 \text{ kmol/s}$$

and

$$W = 0 \cdot 10 - 0 \cdot 0389 = 0 \cdot 0611 \text{ kg/s}$$

From eqn. 7-28

$$q_c/D = (3 \cdot 0 + 1)(27\ 000 - 3340) = 94\ 640 \text{ kJ/kmol}$$

and the condenser duty is

$$94\ 640 \times 0 \cdot 0389 = 3680 \text{ kW}$$

From eqn. 7-21

$$q_r = D \cdot h_D + W \cdot h_w + q_c - F \cdot h_F = (0 \cdot 0389 \times 3340)$$
$$+ (0 \cdot 0611 \times 14\ 700) + 3680 - (0 \cdot 10 \times 10\ 300)$$

Therefore, heat input to reboiler = 3678 kW

(ii) *The feed condition*

To obtain the relationship between L_n and L_m, the liquid flows in the rectifying and stripping sections respectively, a heat balance is taken over the feed plate. If the feed is a liquid at its boiling point, the liquid flowing to the tray below the feed plate is $(L_n + F)$, but if its temperature is below the boiling point, some vapour rising from the plate below will condense to bring the feed to its boiling point. If h_F is the enthalpy of the feed (kJ/kmol), and h_{Fb} the enthalpy of feed at its boiling point (kJ/kmol), the heat required to bring the feed to its boiling point is $F(h_{Fb} - h_F)$ and if λ is the latent heat of vaporization (kJ/kmol), the moles of vapour to be condensed will be $F(h_{Fb} - h_F)/\lambda$.
Then

$$L_m = L_n + F + F(h_{Fb} - h_F)/\lambda = L_n + F(\lambda + h_{Fb} - h_F)/\lambda$$
$$= L_n + q \cdot F \tag{7-28}$$

where q is defined as $\dfrac{\text{heat to vaporize 1 kmol of feed}}{\text{molal latent heat of the feed}}$ (7-29)

The equation of the q line is

$$y_q = x_q \cdot q/(q - 1) - x_F/(q - 1) \tag{7-30}$$

where x_q, y_q are the points of intersection of the rectifying and stripping operating lines. Thus the condition of the feed determines this point of intersection and the slope of the q-line is given by $q/(q - 1)$. The effect of the condition of the feed is shown in Fig. 7-8 and we see that not only will there be a change in the number of plates required depending upon the feed condition, but also an effect on the reboiler heat input if the feed is cold.

(iii) *Graphical constructions*

The methods of determining the number of plates by graphical means are now considered.

Example 7-6: A continuous distillation column is fed with a benzene-toluene mixture at its boiling point containing 35 mol% benzene. The top product is to contain 95 mol% of benzene and the

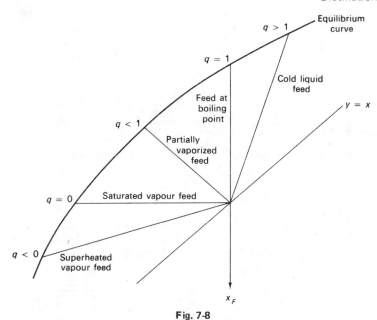

Fig. 7-8

bottoms are not to contain more than 5 mol%. The reflux ratio is to be 3·0 and the number of theoretical plates and the position of the feed plate are to be determined. The following $x - y$ data (mole fraction) are available

x	0	0·1	0·2	0·3	0·4	0·5	0·6	0·7	0·8	0·9	1·0
v	0	0·23	0·38	0·50	0·62	0·70	0·78	0·86	0·92	0·97	1·0

Solution The graphical method of McCabe–Thiele may be applied to this problem, and from the data supplied the equilibrium diagram may be drawn, as shown in Fig. 7-9.

The q-line is vertical for a feed at its boiling point and may be drawn through $x_A = x_F = 0·35$.

The slope of the rectifying operating line is given by $R/R + 1) = 0·75$ or it may be drawn by joining the points (x_D, x_D) and the intercept at $x_A = 0$ at $y_A = x_D/(R + 1) = 0·238$. The stripping operating line is obtained by joining the point of intersection of the q-line with the upper operating line and the point (x_w, x_w).

By stepping off down the curve as shown, we can see that 11 theoretical stages are required or 10 theoretical plates plus the reboiler and condenser. By numbering the plates as shown in Fig. 7-9, the plate

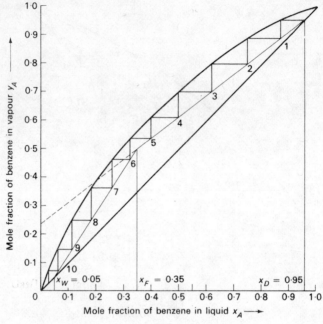

Fig. 7-9

most nearly corresponding to the feed condition is plate 5 and the feed would be deposited on this plate.

Thus, 10 theoretical plates are required, with plate 5 (from the top of the column) being the feed plate.

Example 7-7: Using the data of Example 7-6, show by graphical means how the number of theoretical plates varies with reflux ratio.

Solution The equilibrium data are replotted in Fig. 7-10. The top and bottom section operating lines corresponding to the minimum reflux ratio, R_{min}, are obtained by joining (x_D, x_D) and (x_w, x_w) respectively to the point of intersection of the q-line and the equilibrium curve. In this way, a 'pinched-in' condition exists where an infinite number of plates are required to effect the separation. The minimum reflux ratio is found by producing the rectifying operating line to the y axis and obtaining the intercept as 0.336.

Therefore, as $0.336 = x_D/(R_{min} + 1)$, $R_{min} = 1.83$.

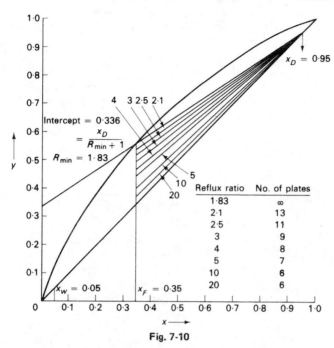

Fig. 7-10

The condition of infinite reflux ratio corresponds to the minimum number of plates and the operating lines become the line $x_A = y_A$, i.e. the diagonal of the diagram. For this case and for all intermediate values of the reflux ratio, the number of theoretical plates is obtained by stepping off as shown in Example 7-6. The steps are omitted for clarity but the resultant figures are plotted in Fig. 7-11.

Minimum reflux with infinite plates and infinite reflux with minimum plates represent the two extreme conditions of continuous distillation. With one the lowest operating costs are combined with infinite capital cost, and with the other, lowest capital cost combined with infinite operating costs. At some point between the two lies a true optimum, and for a rule of thumb approximation an initial selection of the value of the actual reflux ratio may be chosen within the range of $1·1$ to $1·3 R_{min}$.

Example 7-8: Using the equations of Underwood and Fenske, calculate the minimum reflux ratio and the minimum number of plates using the system of Example 7-6, if the relative volatility may be

179

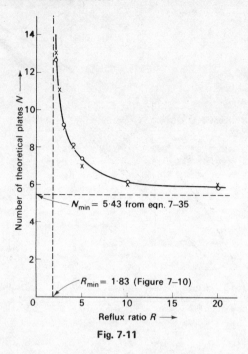

Fig. 7-11

assumed constant at $2 \cdot 5$. Using the correlation of Gilliland, estimate how the number of theoretical plates varies with reflux ratio.

Solution

The minimum reflux ratio R_{min} may be calculated from

$$R_{min} = \frac{1}{\alpha_{ab} - 1} \left[\frac{x_{Da}}{x_{Fa}} - \alpha_{ab} \cdot \frac{x_{Db}}{x_{Fb}} \right] \qquad (7\text{-}31)$$

and for the minimum number of plates N_{min} from

$$N_{min} = \frac{\log_{10} \left[\left(\frac{x_a}{x_b} \right)_D \left(\frac{x_b}{x_a} \right)_W \right]}{\log_{10} \alpha} - 1 \qquad (7\text{-}32)$$

In this case

$$x_{Da} = 0 \cdot 95 \qquad x_{Db} = 0 \cdot 05$$
$$x_{Fa} = 0 \cdot 35 \qquad x_{Fb} = 0 \cdot 65$$

$$x_{wa} = 0.05 \qquad x_{wb} = 0.95$$
$$\alpha = 2.5$$

Hence

$$R_{min} = \frac{1}{2.5 - 1}\left[\left(\frac{0.95}{0.35}\right) - 2.5\left(\frac{0.05}{0.65}\right)\right] = 1.68$$

and

$$N_{min} = \frac{\log_{10}\left(\dfrac{0.95}{0.05}\right)\left(\dfrac{0.95}{0.05}\right)}{\log_{10} 2.5} - 1 = 5.43$$

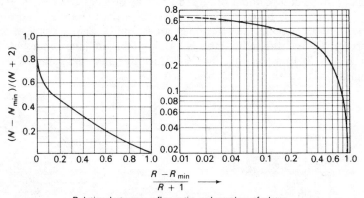

$$\frac{R - R_{min}}{R + 1} \longrightarrow$$

Relation between reflux ratio and number of plates

Fig. 7-12

Gilliland's correlation is presented in Fig. 7-12 and by selecting values of R, corresponding values of $(R - R_{min})/(R + 1)$ may be calculated. The graph is then used to obtain the values of $(N - N_{min})/(N + 2)$ from which the number of plates at the chosen reflux ratios may be calculated.

R	$(R - R_{min})/(R + 1)$	$(N - N_{min})/(N + 2)$	N
2	0.107	0.53	13.8
3	0.330	0.37	9.8
4	0.464	0.29	8.4
5	0.553	0.23	7.4
10	0.756	0.098	6.2
20	0.87	0.060	5.9

The relationship between R and N is included on Fig. 7-12 and it will be seen that close agreement has been obtained with the graphical construction.

(iv) *Multiple feeds and sidestreams*

The simple example of Fig. 7-6 may well be further complicated by the introduction of feed at more than one point and/or by the withdrawal of intermediate product streams. Using the basis of the graphical methods just considered, an example on alternative arrangements should make the necessary procedure clear.

Example 7-9: A continuous distillation column is to separate a mixture of A and B. Two equimolar feed streams F_1 and F_2 enter the column with compositions of 50 and 25 mol% of A respectively. F_1 is a liquid feed at its boiling point and F_2 is a saturated vapour. The top and bottom product specifications are 95 mol% and 5 mol% of A respectively and the reflux ratio is $3 \cdot 0$. Equilibrium data are presented in Fig. 7-13. Construct the McCabe-Thiele diagram and estimate the number of theoretical plates required.

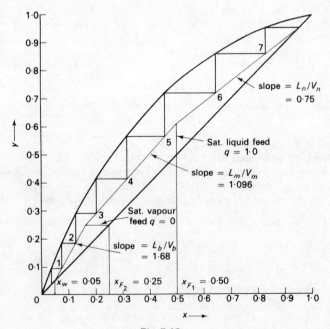

Fig. 7-13

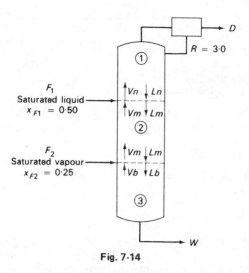

Fig. 7-14

Solution: As a basis for the solution of the problem assume a total feed rate of 100 kmol, i.e. $F_1 = F_2 = 50$ kmol. The system is shown diagrammatically in Fig. 7-14.

An overall mass balance and a balance on component A will establish W and D, thus

$$100 = D + W$$

and

$$(50 \times 0\cdot5) + (50 \times 0\cdot25) = (D \times 0\cdot95) + (W \times 0\cdot05)$$

from which, $D = 36\cdot1$ kmol and $W = 63\cdot9$ kmol

A series of mass balances will determine the liquid and vapour flows in each part of the column.
In Section (1), $L_n/D = 3\cdot0$, hence

$$L_n = 3\cdot0 \times 36\cdot1 = 108\cdot3 \text{ kmol}$$

and

$$V_n = L_n + D = 144\cdot4 \text{ kmol}$$

A balance around the feed point for F_1, remembering that F_1 is a liquid at its boiling point gives

$$L_m = L_n + F_1 = 108\cdot3 + 50\cdot0 = 158\cdot3$$

183

and

$$V_m = V_n + L_m - L_n - F = V_n + L_n + F_1 - L_n - F$$

hence

$$V_m = V_n = 144 \cdot 4 \text{ kmol}$$

A balance around the feed point for F_2, with F_2 being a saturated vapour gives

$$V_m = V_b + F_2 = 94 \cdot 4$$

and

$$F_2 + L_m + V_b = V_m + L_b$$

hence

$$L_b = L_m = 158 \cdot 3 \text{ kmol}$$

To check the mass balances, a balance over Section (3) gives

$$F_2 + L_m = V_m + W$$

From which W is 63:9, which checks the overall balance.

The slope of the operating lines on the McCabe–Thiele diagram is equal to L/V and hence for each section:

Section (1) $L_n/V_n = 108 \cdot 3/144 \cdot 4 = 0 \cdot 75$
  (2) $L_m/V_m = 158 \cdot 3/144 \cdot 4 = 1 \cdot 096$
  (3) $L_b/V_b = 158 \cdot 3/94 \cdot 4 = 1 \cdot 68$

Hence the operating lines may be constructed on the equilibrium diagram (Fig. 7-13) as shown, noting that $q = 1 \cdot 0$ and 0 for F_1 and F_2 respectively, and the number of theoretical plates may be estimated in the usual way. From Fig. 7-13 it is seen that seven theoretical plates are required.

In the case of sidestreams, a similar approach is adopted. If saturated liquid is withdrawn, the slope of the operating line will be reduced; if saturated vapour is removed, the slope will increase.

(v) The Ponchon–Savarit Method

McCabe–Thiele constructions are limited by the assumptions of constant molal overflow and mutual insolubility of the phases. The method of Ponchon and Savarit uses enthalpy–concentration data and to illustrate its use two examples are included.

Example 7-10: A 30 mol% solution of A in B at its boiling point is to be continuously distilled to give a top product of 95% A and a bottom product containing not more than 3% of A. The reflux ratio is to be twice the minimum value and an estimate of the number of theoretical plates required to effect the separation is to be obtained by the Ponchon–Savarit method and compared with the value obtained by the McCabe–Thiele method. In addition, the heat to be removed in the condenser and the heat to be supplied in the reboiler are to be determined.

Equilibrium data

x_A	0	0·05	0·10	0·15	0·20	0·30	0·40	0·50	0·60	0·80
y_A	0	0·30	0·58	0·71	0·79	0·90	0·96	0·98	0·99	0·995

Enthalpy data (kJ/kmol)

x_A	0	0·20	0·40	0·60	0·80	0·9	0·9
Liquid	767	418	198	116	163	209	279
Vapour	2765	2580	2370	2160	1905	1765	1490

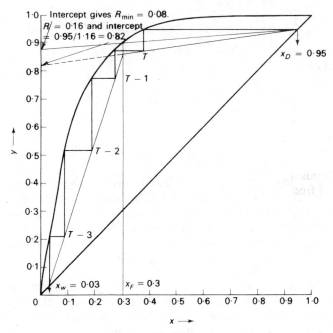

Fig. 7-15

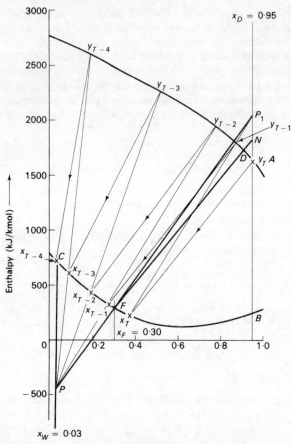

Fig. 7-16

Solution The equilibrium data are plotted in Fig. 7-15 and the enthalpy data in Fig. 7-16, and we will use these figures to outline a summary of the use of the Ponchon–Savarit method. No proofs of the method are presented as these are available in standard texts.

On the enthalpy – composition diagram, vertical lines are constructed at $x_w = 0.03$, $x_F = 0.30$ and $x_D = 0.95$.

The point F is the intersection of the feed composition line and the liquid enthalpy curve. The point N is located by producing the line FD to the vertical through x_D. The point D on the vapour enthalpy curve corresponds to the value of y in equilibrium with $x_F = 0.30$ and is obtained from the equilibrium curve at $y = 0.903$.

The point N enables the minimum reflux ratio to be found from

$$R_{min} = \text{length } NA/\text{length } AB$$

and the length NB represents the heat removed in the condenser per unit mass of liquid at its boiling point. Points vertically above N correspond to actual reflux ratios (such as P_1), and it will be seen that the greater the value of R, the greater is the distance P_1N and fewer plates are required for the separation. However, as P_1B represents the condenser duty, this rises as well, so that an economic selection of the reflux ratio must be made as indicated earlier.

The point P_2 is located by producing the line P_1F to its intersection with the vertical through x_w. The length P_2C represents the heat input to the reboiler per unit mass of bottom product.

The method of obtaining the number of theoretical plates will be explained by means of the problem stated.

Point F is located at the intersection of the feed composition $x_F = 0.30$ and the liquid enthalpy line. From the equilibrium curve x_F is in equilibrium with $y = 0.903$ and hence the point D is located on the vapour enthalpy line. FD is produced to $N = 1850$ kJ/kmol.

Then

$$R_{min} = NA/NB = (1850 - 1640)/(1640 - 250) = 0.151$$

Actual value of reflux ratio is $2 \times 0.151 = 0.302$
Hence P_1 has co-ordinates (0.95, 2060).
Produce P_1F to P_2 at (0.03 − 420).

The vapour leaving the top plate has a composition $y_t = 0.95$. From the $x - y$ data, $x_T =$ liquid on the top plate $= 0.375$ Hence the equilibrium tie line is produced between y_T and x_T. The composition of the vapour rising from plate $T - 1$ may be found from the intersection of the line x_TP_1 and the vapour enthalpy curve at $y_{T-1} = 0.87$ and from the $x - y$ data, $x_{T-1} - 0.27$. This tie line cuts the line P_2FP_1 and from now on the point P_2 is used.

The line P_2x_{T-1} is produced to $y_{T-2} = 0.775$ from which $x_{T-2} = 0.19$. Producing P_2x_{T-2} gives $y_{T-3} = 0.52$ and $x_{T-3} = 0.082$. The procedure is repeated to give $y_{T-4} = 0.185$ and $x_{T-4} = 0.03$ which is the bottoms specification. Thus, x_{T-4} represents the still composition and y_{T-4}, the vapour rising from the reboiler. The number of theoretical plates is seen to be four and the feed is introduced on to the second plate from the top.

The McCabe–Thiele construction of Fig. 7-15 also requires four theoretical plates with the same position for the feed plate.

The heat to be removed in the condenser per unit mass of product is

length $P_1 B$ = 2060 − 250 = 1810 kJ/kmol

Heat required in the reboiler per unit mass of product is

length $P_2 C$ = 720 − (− 420) = 1140 kJ/kmol

The Ponchon–Savarit method makes no assumptions regarding constant molar latent heat and the neglecting of the heat of mixing which are the basis of the McCabe–Thiele method. Thus, for non-ideal systems, the enthalpy − composition data should be used whenever possible. In this example, both methods produced the same answer but the Ponchon method should, in general, be regarded as more reliable. In addition, as has been shown, heat loadings may be simply and accurately calculated.

Example 7-11: A feed stream of A and B consisting of boiling liquid and saturated vapour whose enthalpy is 1250 kJ/kmol contains 42·5 mol% of A, and is to be separated to produce a distillate of 99% of A and a bottom product containing not more than 2% of A. Using the data of the previous example calculate:

   (a) The minimum number of theoretical plates in the column to effect the separation.
   (b) The minimum reflux ratio, R_{min}.
   (c) The number of actual plates required if the column efficiency is 50% and the reflux ratio is $1·2 R_{min}$.
   (d) The optimum feed plate for this condition.

Solution   The enthalpy data are replotted in Fig. 7-17.

The minimum number of plates occurs at total reflux when the point P_1 on $x = x_D$ is at infinity. Hence the operating lines are vertical. The condition of the feed is located between the liquid and vapour enthalpy lines for a partly vaporized feed. (For a feed at its boiling point or as a saturated vapour, F lies on the liquid and vapour enthalpy lines respectively, and for a liquid below its boiling point or as a superheated vapour it lies below and above the respective curves). Thus, with the composition and percent vapour specified, the point F may be located.

   (a) Starting at the top plate, the vapour leaving the plate has a composition $y_T = 0·99$. The corresponding value of x_T from the equilibrium data is $x_T = 0·60$. Hence the equilibrium tie line $y_T x_T$ is drawn. At total reflux, the operating line is vertical and $y_{T-1} = 0·60$ and $x_{T-1} = 0·11$. y_{T-2} is located

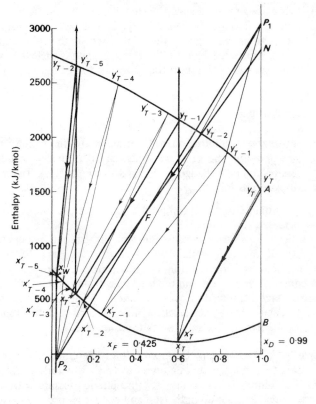

Fig. 7-17

at $y_{T-2} = 0 \cdot 11$ and x_{T-2} is found to equal $x_W = 0 \cdot 02$. Thus two theoretical plates are required at total reflux.

(b) The minimum reflux ratio is located by producing the tie line which passes through F to the line $x = x_D$ at N.
Then

R_{min} = length NA/length NB

$= (2810 - 1520)/(1520 - 290) = 1 \cdot 049$

(c) $R = 1 \cdot 2\, R_{min} = 1 \cdot 259$. Hence the point $P_1 = 5068$ kJ/kmol and $P_2 = -70$ kJ/kmol. The theoretical stages are stepped off as indicated in the previous example and it is found that five ideal plates are required. As the plate efficiency is 50%, the number of actual plates is $5/0 \cdot 50 = 10$ plates.

189

(d) The feed tray is located on the diagram by the tie line which passes through the line P_1FP_2. In this case, the third ideal tray from the top (i.e. the middle tray in the column) meets this condition and with a 50% plate efficiency would correspond to the sixth tray from the top of the column.

7.3.4 Batch distillation

In batch distillation, an initial quantity of material is fed to the still and during the operation one or more phases are continually withdrawn. The process results in a change of composition with time. Two modes of operation are possible:

1) To maintain a constant top product specification, the reflux ratio may be increased continuously.

2) If the reflux ratio remains constant throughout the operation, the quality of the top product will decrease with time. In order to obtain a product of composition x_D, the initial product removed will have a composition $x > x_D$ and the process will stop when $x < x_D$ such that the average composition over the whole operation is equal to x_D.

The McCabe—Thiele construction may be used and the following examples will serve to illustrate the methods involved.

Example 7-12: An equimolal mixture of ethanol and water is to be fractionated in a batch column equivalent to six theoretical plates plus a still. The column is to operate at atmospheric pressure with a total condenser, and the hold-up in the still and condenser may be neglected. Calculate:

(a) The amount of distillate obtained and the heat required per 100 kmol of feed for a top product of 80 mol% alcohol if the reflux ratio is not to be increased beyond 6·0.
(b) The average composition of the distillate and the heat required if the composition of the still is not to fall below 10 mol% alcohol when operating with a constant reflux ratio of 4·0.

Equilibrium Data

x	0	0·05	0·1	0·2	0·3	0·4	0·5	0·6	0·7	0·8	0·89
y	0	0·32	0·44	0·53	0·57	0·61	0·65	0·70	0·75	0·82	0·89

An average latent heat of 40 MJ/kmol may be assumed.

Solution: Assume an initial batch of S_1 kmol of material of composition x_{S1} is fed to the still and a product of D kmol of composition x_D is required. As the still operates, the top product will be initially rich in respect of the more volatile component while the still composition will become progressively weaker. As a result, the product purity will steadily fall. In order to produce a product of specified composition, two alternative methods are available and these are illustrated by the two parts of the question.

(a) In order to maintain a constant value of x_D, the reflux ratio must be increased progressively.

If the initial charge to the still is S_1 moles of composition x_{S1} and if the final quantity remaining in the still is S_2 moles of composition x_{S2}, the initial and final reflux ratios are R_1 and R_2 respectively, the following analysis results.

An overall mass balance gives:

$$S_1 - S_2 = D \qquad (7\text{-}33)$$

A balance on the more volatile component gives:

$$S_1 \cdot x_{S1} - S_2 \cdot x_{S2} = D \cdot x_D \qquad (7\text{-}34)$$

and

$$D = S_1 \left[(x_{S1} - x_{S2})/(x_D - x_{S2}) \right] \qquad (7\text{-}35)$$

If i_1 and i_2 are the intercepts of the y axis of the equilibrium diagram corresponding to the reflux ratios R_1 and R_2

$$i_1 = x_D/(R_1 + 1); \quad i_2 = x_D/(R_2 + 1)$$

Thus the final reflux ratio may be found (if it is not specified) and the total quantity of distillate may be calculated.

In part (a) of the problem to find the quantity of distillate: $S_1 = 100$ kmol, $x_{S1} = 0 \cdot 50$ $x_D = 0 \cdot 80$, x_{S2} may be found by graphical means as the still is equivalent to six plates plus the still and condenser, i.e. seven stages. The equilibrium diagram is presented in Fig. 7-18 and $i_2 = 0 \cdot 80/(R_2 + 1)$. As $R_2 = 6 \cdot 0$, $i_2 = 0 \cdot 114$ and the operating line may be constructed. As seven stages are available, they may be stepped off from the point (x_D, x_D) to give $x_{S2} = 0 \cdot 075$.

From eqn. 7-35

$$D = S_1 \left[(x_{S1} - x_{S2})/(x_D - x_{S2}) \right] = 100[(0 \cdot 50 - 0 \cdot 075)/$$
$$(0 \cdot 80 - 0 \cdot 075)]$$

$$= 58 \cdot 6 \text{ kmol distillate}/100 \text{ kmol charged to the still}$$

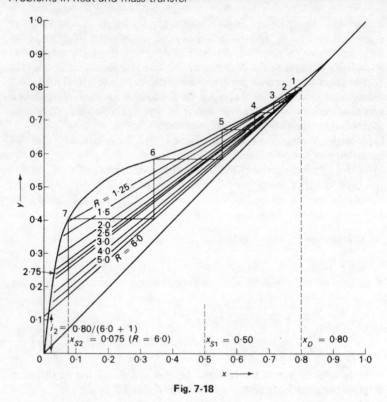

Fig. 7-18

If the reflux ratio R is continuously adjusted to keep x_D constant, the value of R at any moment is given by $R = dL/dD$ where L is the quantity of reflux liquid. Hence, during the operation

$$\int_0^L dL = \int_{R_1}^{R_2} R\,dD \tag{7-36}$$

To provide the reflux dL, an amount of heat $(= \lambda dL)$ must be removed in the condenser where λ is the latent heat/kmol. Thus the heat to be supplied in the boiler over the period of operation is

$$Q_R = \lambda \int_0^L dL = \lambda \int_{R_1}^{R_2} R\,dD \tag{7-37}$$

To find the heat required, it is necessary to obtain the relationship between R and D for the operation by a method similar to that just employed, and by graphical integration the value of $\int_{R_1}^{R_2} R \cdot dD$ may be found. It will be seen that the separation may not be achieved if the reflux ratio is less than approximately 2·6 as shown by the operating lines on the diagram. For values of R greater than 2·75, corresponding values of x_{S2} are obtained as before and values of D calculated as shown in the table below

R	$x_D(R+1)$	x_{S2}	$(x_{S1}-x_{S2})$	(x_D-x_{S2})	D
2·75	0·213	0·475	0·025	0·325	7·60
3·0	0·200	0·450	0·050	0·350	14·3
4·0	0·160	0·300	0·200	0·500	40·0
5·0	0·133	0·090	0·410	0·710	57·7
6·0	0·075	0·075	0·425	0·725	58·6

The area under the plot of R vs D in Fig. 7-19 enables $\int_{R_1}^{R_2} R \cdot dD$ to be evaluated as 95·8 kmol.

Thus the heat to be supplied to heat the reflux is

$$96 \times 40 = 3840 \text{ MJ}/100 \text{ kmol}$$

(b) If the reflux ratio remains constant throughout the operation, the composition of the top product will fall. Over a small interval of time dt the product composition will fall from x_D to $(x_D - dx_D)$. If, in this time, an amount of product dD is obtained, then from mass balances

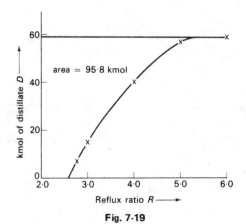

Fig. 7-19

overall and on the more volatile component produces the relationship

$$\ln(S_1/S_2) = \int_{x_{S_2}}^{x_{S_1}} dx_S/(x_D - x_S) \tag{7-38}$$

The integral may be evaluated by plotting $1/(x_D - x_S)$ against x_S (Fig. 7-20) and hence the amount of distillate may be found as $D = S_1 - S_2$. The amount of heat required is then obtained from

$$Q = \lambda RD \tag{7-39}$$

With reference to Fig. 7-20, the gradient of the operating line is constant at $R/(R + 1) = 0\cdot8$. For various distillate compositions, operating lines are drawn and the corresponding values of x_S found by stepping off. The resultant figures are plotted as shown in the inset and the area under the curve for values of x_S between $0\cdot10$ and $0\cdot5$ (the feed composition) is found to be $0\cdot825$.

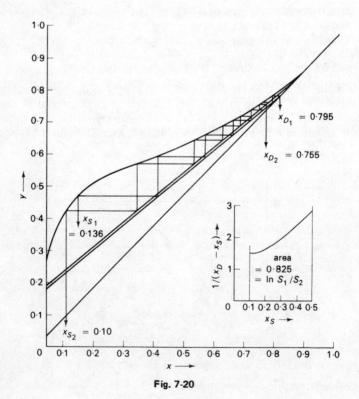

Fig. 7-20

Hence

$$\ln(S_1/S_2) = 0.825$$

and

$$S_1/S_2 = 2.25$$

Thus

$$D = S_1 - S_2 = 100 - 100/2.25 = 55.6 \text{ kmol}/100 \text{ kmol}$$

The average composition may also be calculated.
Amount of alcohol in product is

$$S_1 \cdot x_{S1} - S_2 \cdot x_{S2} = 100(0.5) - (100 - 55.6)(0.10) = 45.6 \text{ kmol}$$

Average composition is

$$45.6/55.6 = 0.82 \text{ mol fraction of alcohol}$$

The heat required is

$$55.6 \times 4.0 \times 40 = 8896 \text{ MJ}/100 \text{ kmol}$$

Although the quality of the product is higher in this case, the heat required for the process is more than twice the earlier figure.

In the previous example, the hold-up in the still and the condenser was neglected. In practice, there is always some finite hold-up on the plates or within the voids of the packing. The following example serves to illustrate the effect that hold-up can have on the recovery of product from the still.

Example 7-13: 200 kmol of an equimolal mixture of A and B are to be distilled in a batch still to give a constant top product composition of 95% of A. The relative volatility is 2.5 and the choice of columns lies between a six plate column with a liquid hold-up of 8.0 kmol/plate and a five plate column with a hold-up of 3.0 kmol/plate. Which column will give the higher recovery of A if the hold-up in the condenser and reflux pipes is neglected?

Solution The $x - y$ equilibrium data are first calculated from

$$y_A = \alpha \cdot x_A /[1 + (\alpha - 1)x_A]$$

Thus

x	0	0.10	0.20	0.30	0.40	0.50	0.60	0.70	0.80	0.90	1.0
y	0	0.22	0.38	0.52	0.63	0.71	0.79	0.85	0.91	0.96	1.0

and this relationship is presented in Fig. 7-21.

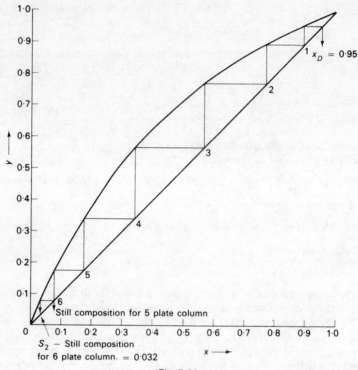

Fig. 7-21

The problem may be simplified by making the assumption that at total reflux, the vapour entering a plate has the same composition as the liquid on that plate. Thus the liquid compositions may be read from the $x - y$ diagram at total reflux with $x_D = 0.95$ and the table opposite produced.

An overall mass balance gives

$$S_1 - S_2 = D + \Sigma h \tag{7-40}$$

where h = hold up on each plate (kmol).
For the more volatile component

$$S_1 \cdot x_{S1} - S_2 \cdot x_{S1} = D \cdot x_D + \Sigma h \cdot x_A \tag{7-41}$$

where $\Sigma h \cdot x_A$ is calculated as shown in the table.

196

Plate (numbered from the top of the column)	Liquid composition x_A	Hold-up in kmol of A($= h \cdot x_A$) 6 plate col. $h = 8 \cdot 0$	5 plate col. $h = 3 \cdot 0$
1	0·890	7·12	2·670
2	0·715	5·72	2·146
3	0·565	4·52	1·696
4	0·390	3·12	1·170
5	0·175	1·40	0·526
6	0·080	0·64	—
		$\Sigma h x_A = 22 \cdot 52$	$\Sigma h x_A = 8 \cdot 208$

For the six plate column:

$$200 - S_2 = D + (6 \times 8)$$

and

$$200(0 \cdot 5) - S_2 (0 \cdot 032) = D(0 \cdot 95) + 22 \cdot 52$$

from which

$$S_2 = 72 \cdot 90$$

and the recovery

$$D = 79 \cdot 1 \text{ kmol}$$

For the five plate column

$$200 - S_2 = D + (5 \times 3 \cdot 0)$$

and

$$200(0 \cdot 5) - S_2(0 \cdot 080) = D(0 \cdot 95) + 8 \cdot 208$$

from which

$$S_2 = 93 \cdot 34 \text{ kmol}$$

and the recovery

$$D = 88 \cdot 66 \text{ kmol.}$$

Thus, the recovery is some 12% greater using fewer of the lower hold-up trays indicating the importance of the hold up consideration.

7.3.5 Multicomponent distillation

Multicomponent distillation refers to the separation of mixtures containing more than two components. In order to obtain n pure components from a mixture of n constituents, $(n-1)$ columns are required and this may be achieved by removing (say) the lightest components from the top of one column and taking the remaining $(n-1)$ components from the bottom and passing them to another column.

If the phase rule is applied to multicomponent systems, it is not possible to specify the compositions of all the components in the product streams. The concept of *key components* is introduced where the light and heavy key components are those whose specifications are fixed in the top and bottom streams. All components lighter than the light key appear mostly in the overhead product and all those heavier than the heavy key appear mostly in the bottoms. The main purpose of the distillation process is the separation of the key components.

In this short section, no attempt is made to cover all the aspects and methods available for the solution of multicomponent distillation problems. Two methods of obtaining the minimum reflux ratio are considered and two methods of calculating the number of plates required for a given separation are illustrated by means of worked examples.

Example 7-14:   A mixture of A, B, C and D at its boiling point is to be continuously distilled to give the top and bottom product specifications in the table below. Calculate the bottoms composition and obtain estimates of the minimum reflux ratio by the approximate method of Colburn and by that of Underwood.

Component	Feed Composition (x_F)	Distillate Composition (x_D)	Bottoms Composition (x_W)	Relative Volatility (α)
A	0·30	0·95		1·25
B	0·15	0·04	0·20	1·00
C	0·30	0·01		0·60
D	0·25	0		0·40

Solution: The composition of the bottom stream may be obtained from a series of mass balances:

Basis 100 kmol of feed.
Overall

$$100 = D + W$$

Mass balance on B

$$100(0\cdot15) = D(0\cdot95) + W(0\cdot20)$$

from which

$$D = 31\cdot25, \quad W = 68\cdot75$$

Mass balance on A

$$100(0\cdot30) = 31\cdot25(0\cdot95) + 68\cdot75(x_{WA})$$

from which

$$x_{WA} = 0\cdot00451$$

Mass balance on C

$$100(0\cdot30) = 31\cdot25(0\cdot01) + 68\cdot75(x_{WC})$$

from which

$$x_{WC} = 0\cdot4318$$

Mass balance on D

$$100(0\cdot25) = 31\cdot25(0) + 68\cdot75(x_{WD})$$

from which

$$x_{WD} = 0\cdot3636$$

and

$$\Sigma x_W = 0\cdot999\ 946 \simeq 1\cdot00$$

Colburn's method for R_{min} is based on Underwood's equation for a binary system using the light and heavy key components.

$$R_{min} = \frac{1}{\alpha_{AB}-1} \left[\frac{x_{DA}}{x_{nA}} - \alpha_{AB} \frac{x_{DB}}{x_{nB}} \right] \tag{7-42}$$

where α_{AB} is the volatility of the light key relative to the heavy key, x_{DA}, x_{nA} the top and pinch compositions of the light key component, and x_{DB}, x_{nB} the top and pinch compositions of the heavy key component.

199

also

$$x_{nA} = \frac{r_f}{(1 + r_f)(1 + \Sigma\alpha \cdot x_{fh})} \tag{7-43}$$

and

$$x_{nB} = x_{nA}/r_f$$

where r_f is the estimated ratio of the key components on the feed plate. (When the feed is a liquid at its boiling point, r_f equals the ratio of the key components in the feed, otherwise it is the ratio of the key components in the liquid part of the feed.) x_{fh} is the mole fraction of each component in the liquid part of the feed heavier than the heavy key, and α the volatility of the component relative to the heavy key.

In this example, component A is the light key and B is the heavy key, and $\alpha_{AB} = 1\cdot25$ and $r_f = 0\cdot30/0\cdot15 = 2\cdot0$, thus

$$\Sigma\alpha \cdot x_{fh} = (0\cdot6 \times 0\cdot3) + (0\cdot4 \times 0\cdot25) = 0\cdot28$$

Hence

$$x_{nA} = \frac{2}{(1 + 2)(1 + 0\cdot28)} = 0\cdot5208$$

and

$$x_{nB} = 0\cdot5208/2 = 0\cdot2604$$

Substituting in the equation for R_{min} gives

$$R_{min} = \frac{1}{1\cdot25 - 1}\left[\frac{0\cdot95}{0\cdot5208} - 1\cdot25 \cdot \frac{0\cdot04}{0\cdot2625}\right] = 6\cdot534$$

This is the quick approximation to R_{min} from Colburn's method. Correction factors are available to enable the value obtained to be checked.

The Underwood equations may be written as follows

$$\frac{\alpha_A x_{FA}}{\alpha_A - \theta} + \frac{\alpha_B x_{FB}}{\alpha_B - \theta} + \ldots \frac{\alpha_n x_{Fn}}{\alpha_n - \theta} = 1 - q \tag{7-44}$$

and

$$\frac{\alpha_A x_{DA}}{\alpha_A - \theta} + \frac{\alpha_B x_{DB}}{\alpha_B - \theta} + \ldots \frac{\alpha_n x_{Dn}}{\alpha_n - \theta} = R_{min} + 1 \tag{7-45}$$

where θ is the root of eqn. 7-44 and has a value between the relative volatilities of the key components, and q the thermal condition of the feed. In this example $q = 1$ as the feed is at its boiling point, and hence

$$\frac{1\cdot25 \times 0\cdot30}{1\cdot25 - \theta} + \frac{1\cdot0 \times 0\cdot15}{1\cdot0 - \theta} + \frac{0\cdot6 \times 0\cdot3}{0\cdot6 - \theta} + \frac{0\cdot25 \times 0\cdot25}{0\cdot4 - \theta} = 0$$

This is solved by trial and error for θ. When $\theta = 1\cdot10$, $1\cdot08$ and $1\cdot084$ the left hand side becomes equal to $+0\cdot551$, $-0\cdot136$ and $+0\cdot010$ respectively and hence $\theta = 1\cdot084$ is taken as the required root. Substitution of θ into eqn. 7-45 gives the required value of the minimum reflux ratio

$$R_{min} = 5\cdot76$$

The following example illustrates how 'K values' are used to obtain the compositions of liquid and vapour existing above a specified tray composition. The method can be extended in a repetitive manner to obtain the number of theoretical trays to effect a particular separation.

Example 7-15: In a multicomponent distillation, the compositions of the feed, product, and bottom streams are as shown below (mol %). The composition of the liquid on the nth tray from the top of column is also indicated and you are required to calculate the composition of the liquid on tray $n - 1$. The K values may be assumed constant and the reflux ratio is equal to $4\cdot0$.

Component	Feed composition	Distillate composition	Bottoms composition	K	Liquid composition on tray n
C_1	0·50	1·1	—	14·0	0·001
C_2	5·8	15·1	—	3·4	1·35
C_3	33·7	83·8	2·3	1·35	60·8
C_4	47·1	—	76·9	0·56	25·4
C_5	12·9	—	20·8	0·70	12·4

Solution   As the liquid composition of C_3 on plate n is between the compositions of the feed and the distillate, this tray is in the rectification section and the operating line for any component is given by

$$y_{n-1} = x_n R/(R + 1) + x_D/(R + 1)$$

Therefore the following operating lines may be obtained

$$C_1 \; y_{n-1} = \frac{4\cdot0}{5\cdot0} x_n + \frac{0\cdot011}{5\cdot0} = 0\cdot8\,x_n + 0\cdot0022$$

$$C_2 \qquad\qquad = 0\cdot8\,x_n + 0\cdot0302$$

$$C_3 \qquad\qquad = 0\cdot8\,x_n + 0\cdot1676$$

$$C_4 \qquad\qquad = 0\cdot8\,x_n$$

$$C_5 \qquad\qquad = 0\cdot8\,x_n$$

Using the above equations, the compositions of the vapours rising to tray n i.e. y_{n-1} can be calculated as

Component	y_{n-1}
C_1	0·002 028
C_2	0·0410
C_3	0·6541
C_4	0·2030
C_5	0·0992

$\Sigma y = 0·999\ 328$ which is sufficiently close to $1·00$ to be acceptable. The composition of the components of the liquid on tray n i.e. x_{n-1} may be calculated from the K values since $y = Kx$ and $x = y/K$, hence

Component	y	K	x_{n-1}
C_1	0·002 028	14·0	0·000 144 9
C_2	0·0410	3·4	0·012 05
C_3	0·6541	1:35	0·484 52
C_4	0·2030	0·56	0·362 50
C_5	0·0992	0·70	0·141 71
			$\Sigma x = 1·000\ 92$

The value of Σx is sufficiently close to $1·00$ to be acceptable and the values of x_{n-1} are the required compositions of the liquid on tray $n - 1$.

The last example in this chapter is designed to show as concisely as possible the method of approach to a multicomponent distillation problem when vapour pressure data are available for the components of the mixture.

Example 7-16: A mixture of A, B, and C at its boiling point is to be continuously distilled in a plate column where the pressure drop across each plate is $5·0$ kN/m^2 and the pressure in the still is 105 kN/m^2. The reflux ratio is $1·5$ and the composition of the feed, distillate and bottoms streams is given in the table below. The vapour pressures of the pure components are also given over a limited range of temperatures as indicated. Calculate the following:

(a) The temperature of the liquid in the reboiler.
(b) The composition of the vapour above the liquid in the reboiler.
(c) The composition of the liquid on the first plate above the reboiler i.e. plate $s + 1$.

(d) The composition of the vapour rising from plate $s + 1$.

(e) The composition of the liquid on plate $s + 2$.

Indicate how the number of theoretical plates in the column would be estimated.

Component	Feed	Compositions (mole fraction) Distillate	Bottoms
A	0·50	0·90	0·05
B	0·36	0·10	0·65
C	0·04	–	0·30

Temperature K	386	388	390	391	393
Component	Vapour pressure (kN/m²)				
A	159	253	279	285	295
B	107	113	119	123	132
C	48·7	51·3	54·0	56·0	60·0

Solution: These data are plotted in Fig. 7-22.

(a) *Calculation of reboiler temperature*

The bottom pressure is 105 kN/m². If the three component mixture can be considered ideal, then the sum of the partial pressures of each component should equal the total pressure, that is

$$\Sigma x \cdot P^\circ = 105 \text{ kN/m}^2$$

By calculating $x \cdot P^\circ$ at each temperature for each component the following table may be produced.

Temperature K	386	388	390	391	393
$0 \cdot 05 \, P_A^\circ$	7·95	12·65	13·95	14·25	14·75
$0 \cdot 65 \, P_B^\circ$	69·55	73·35	77·35	79·95	85·80
$0 \cdot 30 \, P_C^\circ$	14·61	15·39	16·20	16·80	18·00
$\Sigma x \cdot P^\circ$	92·11	101·49	107·50	111·00	118·55

A plot of $x \cdot P^\circ$ vs T is included on Fig. 7-22 and the temperature where $\Sigma x \cdot P^\circ = 105$ kN/m² is 389·25 K.

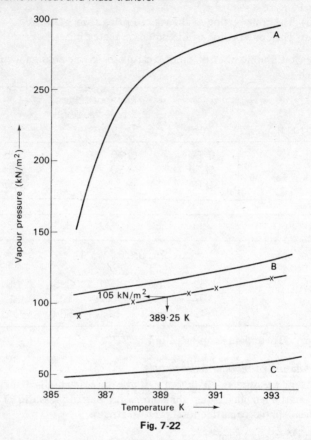

Fig. 7-22

(b) *Calculation of the vapour composition rising from the reboiler*

For each component, $y = \dfrac{x \cdot P^\circ}{P}$

At 389·25,

$$P_A^\circ = 272, \quad P_B^\circ = 117, \quad P_C^\circ = 53 \cdot 0 \ \text{kN/m}^2.$$

Hence

$$y_A = 0 \cdot 130 \quad y_B = 0 \cdot 724 \quad y_C = 0 \cdot 151 \text{ and } \Sigma y = 1 \cdot 005$$

As Σy should equal 1·000, these values are each corrected by the factor $1/1 \cdot 005$ to give

$$y_A = 0\cdot129, \quad y_B = 0\cdot721, \quad y_C = 0\cdot150$$

(c) *Liquid composition on plate*

In order to obtain the liquid compositions, the operating lines must be obtained. Using the usual nomenclature

$$L_n/D = 1\cdot5$$
$$V_n = L_n + D$$

From a mass balance on the basis of 100 kmol of feed, $D = 52\cdot94$ and $W = 47\cdot06$

Hence

$$L_n = 79\cdot41, \quad V_n = 132\cdot35$$

and

$$y_n = (79\cdot41/132\cdot35)x_{n+1} + (52\cdot94/132\cdot95)x_D$$

or

$$y_n = 0\cdot6\,x_{n+1} + 0\cdot4\,x_D$$

In the stripping section

$$V_m = L_m - W \text{ and } L_m = L_n + F = 179\cdot41$$

and

$$V_m = 132\cdot35$$

Hence

$$y_m = 1\cdot356\,x_{m+1} - 0\cdot356\,x_w$$

For component A, $y_{mA} = 1\cdot356\,x_{m+1} - 0\cdot0178$

 B, $y_{mB} = 1\cdot356\,x_{m+1} - 0\cdot231$

 C, $y_{mC} = 1\cdot356\,x_{m+1} - 0\cdot107$

Given y_s, x_{s+1} can be found from these operating lines as

$$x_{s+1}A = 0\cdot108 \quad x_{s+1}B = 0\cdot702 \quad x_{s+1}C = 0\cdot190 \quad \Sigma x_{s+1} = 1\cdot000$$

(d) *Vapour composition above plate s + 1*

The pressure above plate $s + 1$ will be $105 - 5\cdot0 = 100$ kN/m^2 and a trial and error procedure must be adopted to find the temperature using the fact that at the correct temperature, $\Sigma x \cdot P^\circ/P = 1\cdot00$. The following table may be computed.

	P°	T = 386 K x_{s+1}	$y_{s+1} = x \cdot P^\circ /P$	T = 385·9 K P°	y_{s+1}
A	158	0·108	0·171	150	0·162
B	107	0·702	0·751	106·5	0·748
C	48:5	0·190	0·092	49·5	0·094
			$\Sigma y = 1·014$		$\Sigma y = 1·004$

The values obtained for y_{s+1} are considered accurate enough.

(e) *Liquid composition on plate s + 2*

As in part (3) of this example, the liquid composition may be calculated from the operating lines for each component to give

$$x_{s+2}A = 0·132, \quad x_{s+2}B = 0·722, \quad x_{s+2}C = 0·148. \quad \Sigma x = 1·002$$

Number of theoretical plates

Proceeding up the column, plate by plate, the foregoing calculations may be repeated. Components A and B may be regarded as the light and heavy keys and their ratios in the feed = $0·50/0·36 = 1·38$. In the reboiler, and on plates $s + 1$ and $s + 2$ the ratio A/B is equal to $0·076$, $0·154$ and $0·183$ respectively. This ratio will increase up the column until the feed ratio of $1·38$ is reached. This represents the limit of the stripping section. Above the feed plate, the procedure is repeated in the same way with the difference that in calculating the liquid compositions, the relevant rectifying operating lines are used. The top plate is reached when the compositions of the vapour are equal to the product specifications.

List of symbols

D	quantity of distillate	kmol/s, kg/s, kmol.
F	feed rate	kmol/s, kg/s, kmol.
h	enthalpy of liquid	kJ/kmol, kJ/kg
h	hold up in Example 7-13	kmol
H	enthalpy of vapour	kJ/kmol, kJ/kg
K	"K value"	—
L	liquid flow rate	kmol/s, kmol
N	number of plates	—
P	pressure, partial pressure	kN/m^2
P°	vapour pressure of pure component	kN/m^2

q	ratio, heat to vaporise 1 kmol of feed/molal latent heat of feed	kJ/kJ
q_b	heat added in reboiler	kW
q_c	heat removed in condenser	kW
r_f	ratio of key components at feed plate	–
R	reflux ratio	–
S	quantity in still	kmol
T	temperature	K
V	vapour flow rate	kmol/s, kmol
W	bottoms product rate	kmol/s, kmol
x	liquid composition	–
y	vapour composition	–
α	relative volatility	–
γ	activity coefficient	–
θ	root of eqn. 7-47	–

Subscripts

a, b	components a and b
A, B	components A and B
1, 2	components 1 and 2
D	distillate
F	feed
m, n, b	sections in column
min	minimum
q	q-line
s	still

Further reading

COULSON, J. M., RICHARDSON, J. F., BACKHURST, J. R. & HARKER, J. H. *"Chemical Engineering"* Vol. 2. 3rd Ed. Pergamon Press 1979.

PERRY, J. H. Ed. *"Chemical Engineers Handbook"* 4th Ed. McGraw-Hill Book Co. New York 1963.

TREYBAL, R. E. *"Mass Transfer Operations"* 2nd Ed. McGraw-Hill Book Co. New York 1963.

HENLEY, E. J. & STAFFIN, H. K. "Process Design", John Wiley & Sons New York 1963.

KING, J. C. *"Separation Processes"* McGraw-Hill Book Co. New York 1971.

WEISSBERGER, A. Ed. *"Techniques of Organic Chemistry"* Vol. 4 — Distillation Interscience New York 1951.
ROBINSON, C. S. & GILLILAND, E. R. *"Elements of Fractional Distillation"* 4th Ed. McGraw-Hill Book Co. New York.

Problem 7-1: A binary mixture of B and C is to be separated by continuous rectification at a pressure of $100 \, kN/m^2$. Using the data presented in Fig. 7-21, construct the equilibrium curve for the system for liquid compositions of B ranging from $0 \cdot 5$ to $1 \cdot 0$ mol fraction.

Problem 7-2: If, in the statement of the problem of Example 7-3, the product specification is amended to a value of 15 mol % of A, calculate the quantity of product formed if the other conditions remain unaltered.

$(10 \cdot 7 \, kmol)$

Problem 7;3: Benzene and toluene are to be separated in a continuous distillation column with a feed containing 36 mol % of benzene. The feed is a liquid at its boiling point and the reflux ratio is $3 \cdot 2$. The top and bottom specifications are 95% and 5% of benzene respectively. Estimate graphically the number of theoretical plates required and the position of the feed plate. The equilibrium data from Example 7-6 may be used:

(10, 5th).

Problem 7-4: Derive the Underwood and Fenske equations for minimum reflux ratio and minimum number of plates to effect a given separation. Using these equations, derive the relationship between reflux ratio and the number of theoretical plates using the data of Problem 7-3. Compare the results with those obtained by graphical means.

Problem 7-5: The feed to an atmospheric batch still consists of 50 knol of a mixture of A and B. The feed composition is 25 weight % of A and a constant composition distillate containing 90 weight % of A is required from the still which consists of 6 theoretical plates. The maximum boil-up rate is limited to $0 \cdot 0028 \, kmol/s$. Derive the relationship between time and the quantity of distillate produced. The molecular weights of A and B are 76 and 152 kg/kmol respectively.

Equilibrium Data							
x weight %	A	10	20	40	60	80	100
y weight %	A	5	11	25	43	67	100

Problem 7-6: The system of Problem 7-5 is to be distilled with a constant reflux ratio of 4·0 until the still contains 4 weight % of A. Calculate the quantity of distillate obtained, its average composition and the vapour boil-up.

(18·9 kmol; 0·93; 80 kmol)

Problem 7-7: An ammonia-water mixture, flow 0·282 kg/s, is fed to a continuous distillation column. The feed, top and bottom compositions are 25·9%, 99·4% and 9·5 mol % respectively and the reflux ratio is 8% greater than the minimum reflux ratio. If the overall column efficiency is 60%, how many plates are required for the separation and what is the loading of the condenser and reboiler? The enthalpy and equilibrium data of Example 7-10 may be used.

(17 plates; 5·8 kW; 10·7 kW)

Problem 7-8: In a multicomponent distillation the feed, distillate and bottoms compositions are as shown below.

Component	Composition (mol %)			K value	Liquid composition on tray $(T - 7)$ (mole fraction)
	Feed	Top	Bottom		
C_1	0·5	1·1	—	14·0	0·000 01
C_2	5·8	15·1	—	3·4	0·0135
C_3	33·7	83·8	2·3	1·35	0·608
nC_4	47·1	—	76·9	0·56	0·254
iC_4	12·9	—	20·8	0·70	0·124

If the reflux ratio = 4·0, calculate the composition of the liquid on tray $(T - 8)$.

(0·000 144 8; 0·012 06; 0·4845; 0·3625; 0·1416)

Problem 7-9: A three component mixture is being distilled in the recifying section of a fractionating column. The liquid composition on three consecutive plates is shown below.

Component	Relative Volatility	Liquid composition (mol %)		
		Plate n	Plate $(n + 1)$	Plate $(n + 2)$
1	1·00	31·0	20·0	10·6
2	2·07	48·0	54·0	51·4
3	8·57	21·0	26·0	38·0

If the plates may be assumed 100% efficient, find whether or not the ratio of liquid to vapour flows changes as the column is ascended, and if so, in what way?

(L/V increases up the column).

8. Gas Absorption

8.1 Introduction

Gas absorption may be defined as the removal of one or more selected components by absorption into a suitable liquid. Thus, water may be used to recover acetone (for example) from an air-acetone mixture or concentrated sodium hydroxide may be used to absorb water vapour from moist air. Both these examples are physical processes, though the latter is accompanied by large heat effects whilst the former is normally regarded as a typical isothermal operation. These processes are considered in this chapter by means of worked examples but the other class of gas absorption process, i.e. absorption accompanied by chemical reaction, is not considered since design for this type of application is frequently tied to pilot-plant studies from which the process may be scaled up.

8.2 Equilibrium solubility of gases in liquids

The equilibrium characteristics of gas-liquid systems determine the rate at which a gas dissolves in a liquid. A classification into ideal and non-ideal liquid systems is usually made.

8.2.1 Ideal liquid solutions

When the gas mixture in contact with an ideal liquid solution follows the ideal gas laws, Raoult's Law may be used to obtain the relationship between the partial pressure at equilibrium P_e, its mole fraction in the solution x and its vapour pressure P°, that is

$$P_e = P^{\circ} \cdot x \qquad (8\text{-}1)$$

Raoult's Law applies to both binary and multicomponent systems, and in cases where the gas mixture does not follow the ideal gas laws, the fugacity may be substituted for the pressure terms in eqn. 8-1.

Example 8-1: A gas mixture containing 65% A, 25% B, 8% C and 2% of D is in equilibrium with a liquid at 350 K and 300 kN/m². If the vapour pressure of the pure components at 350 K are 1000, 500, 425 and 100 kN/m² respectively, calculate the liquid composition at equilibrium.

Solution The equilibrium partial pressure P_e is obtained from the product of the volume fraction (i.e. the mole fraction) and the total pressure.

Thus, for component A

$$P_{eA} = 0 \cdot 65 \times 300 = 195 \text{ kN/m}^2$$

similarly

$$P_{eB} = 0 \cdot 25 \times 300 = \quad 75 \text{ kN/m}^2$$
$$P_{eC} = 0 \cdot 08 \times 300 = \quad 24 \text{ kN/m}^2$$
$$P_{eD} = 0 \cdot 02 \times 300 = \quad \underline{6 \text{ kN/m}^2}$$
$$\Sigma P_e = 300 \text{ kN/m}^2$$

The mole fraction of each component in the liquid is obtained from eqn. 8-1 as $x = P_e/P^\circ$, hence

$$x_A = 195/1000 = 0 \cdot 195$$
$$x_B = \quad 75/500 \quad = 0 \cdot 150$$
$$x_C = 24/425 \quad = 0 \cdot 056$$
$$x_D = 6/100 \quad = \underline{0 \cdot 060}$$
$$\Sigma x = 0 \cdot 461$$

The remaining liquid has the composition $1 \cdot 0 - 0 \cdot 461 = 0 \cdot 539$ mole fraction.

8.2.2 Non-ideal liquid solutions

Henry's Law is used where the liquid solution is non-ideal and is often expressed as

$$P_e = H \cdot x \qquad (8\text{-}2)$$

or

$$y_e = H'x \qquad (8\text{-}3)$$

where

$$H' = H/P_T$$

For a wide range of gases, Henry's Law holds where the partial pressure of the gas does not exceed 100 kN/m^2. At higher pressures, values of H apply over a narrower range of partial pressure and the partial pressure of the solute gas as well as the temperature and H, must be specified.

8.3 Mass balances in gas absorption

Consider the schematic diagram of an absorber (or stripper) shown in Fig. 8-1 where the internals may be plates or packing, or it may be a spray tower. Gas-liquid contact is counter-current and the following relationships may be derived:

$$Y = y/(1 - y) = P/(P_T - P) \qquad (8\text{-}4)$$

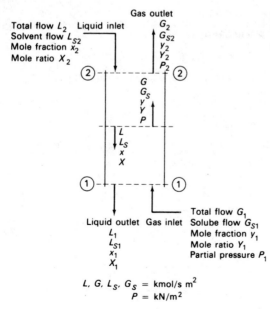

Gas outlet

Total flow L_2 Liquid inlet $\quad G_2$
Solvent flow L_{S2} $\qquad\qquad G_{S2}$
Mole fraction x_2 $\qquad\qquad y_2$
Mole ratio X_2 $\qquad\qquad Y_2$
$\qquad\qquad\qquad\qquad\qquad P_2$

G
G_S
y
Y
P

L
L_S
x
X

Total flow G_1
Liquid outlet Gas inlet Solube flow G_{S1}
L_1 $\qquad\qquad$ Mole fraction y_1
L_{S1} $\qquad\qquad$ Mole ratio Y_1
x_1 $\qquad\qquad$ Partial pressure P_1
X_1

$L, G, L_S, G_S = \text{kmol/s m}^2$
$P = \text{kN/m}^2$

Fig. 8-1

213

Problems in heat and mass transfer

where P_T is total pressure (kN/m^2), and

$$X = x/(1 - x) \tag{8-5}$$
$$G_s = G(1 - y) = G/(1 + Y) \tag{8-6}$$
$$L_s = L(1 - x) = L/(1 + X) \tag{8-7}$$

A mass balance over the lower part of the tower may be expressed in terms of mole ratios or mole fractions as

$$G_s(Y_1 - Y) = L_s(X_1 - X) \tag{8-8}$$

or

$$G(y_1 - y) = L(x_1 - x) \tag{8-9}$$

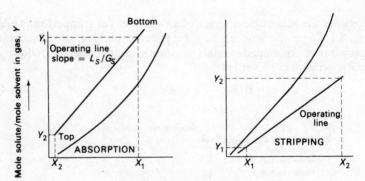

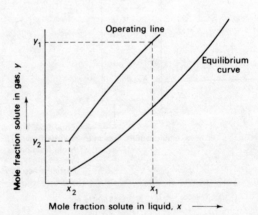

Fig. 8-2

214

or in terms of the solvent and solute flows and mole fractions:

$$G_s\left(\frac{y_1}{1-y_1} - \frac{y}{1-y}\right) = G_s\left(\frac{P_1}{P_T - P_1} - \frac{P}{P_T - P}\right)$$

$$= L_s \frac{x_1}{1-x_1} - \frac{x}{1-x_1} \qquad (8\text{-}10)$$

Equations 8-8 to 8-10 are the equations of the operating line and relate the liquid and gas concentrations in any part of the tower. In absorption processes, the operating line is above the equilibrium curve, while for stripping operations it lies below the equilibrium relationship. It is convenient to plot the operating line in terms of the solvent and solute using mole ratios, since the resulting operating line is always straight. When mole fractions are used, the line is curved, though for dilute compositions $x \triangleq X$ and $y \triangleq Y$ and the operating line is virtually straight. Typical operating lines are shown in Fig. 8-2 and their derivation and application are shown later in this chapter.

8.4 Design calculations

In the design of gas absorption (or stripping) equipment, the gas flow rate, the gas composition, and the operating pressure are normally specified from process conditions. The pressure drop across the absorber, the degree of recovery, and the choice of solvent are sometimes specified but often left to the discretion of the designer.

The design engineer will have to decide upon the column internals, the vessel diameter and height of internals, the selectio of an economical gas/liquid ratio, and have to consider heat effects and whether internal or external heat transfer means should be provided. The following sections and examples deal with some of these factors.

8.4.1 Choice of solvent

The solubility of the gas in the selected solvent should be high to increase the rate of absorption. It should, however, have a low vapour pressure to prevent losses in the exit gas stream. Low viscosity enhances absorption rates, improves flooding characteristics in packed towers and lowers pumping costs. The solvent should be cheap, non-corrosive, non-toxic and non-flammable. Water is commonly used as a solvent and oils are used for the absorption of light hydrocarbons.

8.4.2 Calculation of gas-liquid ratio

The minimum liquid rate is calculated from the entering gas composition and its solubility in the exit liquor, assuming saturation. A choice (if it is not already specified) of the exit gas composition must be made before performing the calculation, and this is usually based on an economic balance between the loss that can be tolerated and the extra cost of a higher tower to obtain high recovery rates.

Example 8-2: Benzene is to be absorbed from a benzene-air mixture by the use of a non-volatile hydrocarbon oil. The inlet gas contains 2.5% by volume of benzene and the exit concentration is to be 0.01%. Operation of the packed tower is to take place at atmospheric pressure and the total gas feed rate is 0.2 kg/s. If the solvent oil enters the tower solvent-free, calculate the minimum oil flow necessary to effect the absorption duty and the exit composition of the oil. The system may be assumed ideal. The vapour pressure of benzene at the mean tower temperature is 13.33 kN/m^2 and the molecular weights of air, benzene and the oil are 29, 78 and 250 kg/kmol respectively.

Solution The system is illustrated in Fig. 8-3.
Let the mass flow of benzene be a kg/s, then the mass flow of air is $(0.2 - a)$ kg/s.

Now, the inlet gas composition is 0.025, then

$$0.025 = (a/78)/[(a/78) + (0.2 - a)/29]$$

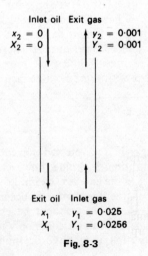

Inlet oil Exit gas
$x_2 = 0$ $y_2 = 0.001$
$X_2 = 0$ $Y_2 = 0.001$

Exit oil   Inlet gas
x_1 $y_1 = 0.025$
X_1 $Y_1 = 0.0256$

Fig. 8-3

from which the mass flows of benzene and air are 0.014 and 0.186 kg/s respectively.

Hence molar flows of benzene and air are $0.000\ 18$ and $0.006\ 41$ kmol/s.

The mole ratio, i.e. kmol benzene/kmol air, at inlet and exit are obtained from $y = Y/(1 + Y)$ where y = mole fraction to give $Y_1 = 0.0256$ and $Y_2 = 0.0001$ respectively.

A mass balance on the benzene over the column gives

$$L(X_1 - X_2) = G(Y_1 - Y_2) \tag{8-8}$$

where L, G = molar flows of oil and air respectively.

The minimum oil flow L_{min} occurs when the exit liquid stream is in equilibrium with the inlet gas, i.e. when x_1 and y_1 are in equilibrium (Fig. 8-4)

$$y_1 = 0.025 = (13.33/101.3)x_1$$

from which

$$x_1 = 0.19 \text{ and } X_1 = 0.235$$

and

$$X_2 = 0 \text{ (solute free)}$$

Hence

$$L_{min}(0.235 - 0) = 0.006\ 41(0.0256 - 0.0001)$$

from which

$$L_{min} = 0.0007 \text{ kmol/s} = 0.175 \text{ kg/s}$$

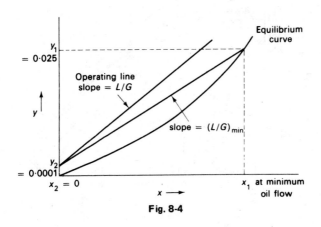

Fig. 8-4

Fig. 8-4 shows the significance of eqn. 8-8 in relation to the value of $(L/G)_{min}$ and the actual operating line of slope $(L/G) > (L/G)_{min}$. The actual liquid-gas ratio is greater than the minimum by 10−100% and this choice of the ratio will result from a cost-optimization on the system.

8.4.3 Selection of column internals

Packed columns are preferred for small diameter columns (<1 m), for corrosive duties, for low pressure drop and for foaming systems. Plate columns are usually employed for diameters greater than 1 m, for low liquid flow rates and where internal cooling is required. These are the general guidelines adopted but each system should be considered individually.

8.4.4 Column diameter

(a) *Packed towers* The diameter of a packed tower is chosen such that the design vapour velocities are between 60 − 80% of the flooding velocity. In the case of vacuum operation, the pressure drop is usually the controlling factor and for both applications, and for all types of packing, the diameter is most conveniently found from the generalized pressure drop correlation presented in Fig. 8-5.

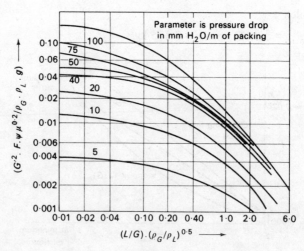

Fig. 8-5

Example 8-3: The average flow conditions existing in a tower packed with 38 mm metal Pall Rings are as follows:

Vapour		*Liquid*	
Molal flow	$G_m = 0{\cdot}01$ kmol/s	$L_m = 0{\cdot}005$ kmol/s	
Density	$\rho_G = 4{\cdot}23$ kg/m^3	$\rho_L = 1036$ kg/m^3	
Molecular mass M_G = 100 kg/kmol		M_L = 150 kg/kmol	
Viscosity	–	$\mu_L = 1{\cdot}61 \times 10^{-3}$ N.s/m^2.	

The packing factor F for the packing is 28 and the allowable pressure drop is 40 mm H$_2$O/m of packing. Estimate the column diameter.

Solution   The generalized pressure drop correlation is shown in Fig. 8-5, and the ordinate and abscissa are first evaluated

$$(L/G)(\rho_V/\rho_L)^{0{\cdot}5} = \left(\frac{0{\cdot}005 \times 150}{0{\cdot}01 \times 100}\right)\left(\frac{4{\cdot}23}{1036}\right)^{0{\cdot}5} = 0{\cdot}048$$

and

$$G'^2 \cdot F \cdot \psi \cdot \mu^{0{\cdot}2}/\rho_G \cdot \rho_L \cdot g$$

$$= G'^2 \cdot 28 \cdot \frac{1036}{1000}(1{\cdot}61 \times 10^{-3})^{0{\cdot}2}/1000(4{\cdot}23 \times 1036 \times 9{\cdot}81)$$

$$= 0{\cdot}000\ 174\ G'^2 \qquad \text{where } G' \text{ is the vapour flow (kg/s m}^2\text{)}.$$

At a pressure drop of 40 mm/m and $(L/G)(\rho_G/\rho_L)^{0{\cdot}5} = 0{\cdot}048$,

$$0{\cdot}000\ 174 G'^2 = 15{\cdot}2 \text{ kg/s m}^2$$

Hence column area is

$$\frac{0{\cdot}01 \times 100}{15{\cdot}2} = 0{\cdot}066 \text{ m}^2$$

and the diameter is $0{\cdot}29$ m.

The diameter obtained is dependent upon the allowable pressure drop and careful consideration should be given to this factor to ensure that the most economical tower is designed.

(b) *Plate towers*   Correlations exist relating to the flooding velocity of the vapour to the flow conditions existing within the tower for most types of tray. The relevant correlation for sieve trays is presented as Fig. 8-6 and its use will be illustrated by means of Example 8-4.

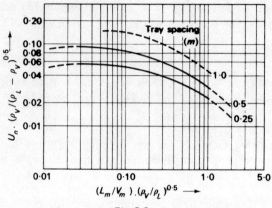

Fig. 8-6

Example 8-4: A sieve tray tower is being considered for the duty defined in Example 8-3. If the tower is to be run at 80% of the flooding velocity, estimate the diameter of the column.

Solution
In this case

$$(L_m/V_m)(\rho_V/\rho_L)^{0.5} = \frac{0.005}{0.01}\left(\frac{4.23}{1036}\right)^{0.5} = 0.032$$

and

$$U_n(\rho_V/\rho_L - \rho_V)^{0.5} = U_n\left(\frac{4.34}{1036 - 4.23}\right)^{0.5} = 0.0641\,U_n$$

A tray spacing must be assumed at this stage and for this duty, a preliminary estimate of 0.5 m would be appropriate. Subsequent design calculations will show whether this is satisfactory or if it may be reduced.

For a tray spacing of 0.5 m and $(L_n/V_m)(\rho_V/\rho_L)^{0.5} = 0.032$.

$$0.0641\,U_n = 0.09$$

from which

$$U_n = 1.40 \text{ m/s}$$

At 80% flooding, the vapour velocity is

$$1.40 \times 0.80 = 1.12 \text{ m/s}$$

The volumetric vapour flow is

$$0.01 \times 100/4.23 = 0.236 \text{ m}^3/\text{s}$$

Hence the tower area is

$$0.236/1.12 = 0.211 \text{ m}^2$$

and the diameter is 0.52 m.

Examples 8-3 and 8-4 show how the resulting tower diameter depends upon the percentage flooding permitted and the specified pressure drop. The packing factor F is determined by the type of packing and assumes larger values for decreasing packing size and for less efficient packings. Thus, if 38 mm metal Raschig Rings were to be used in Examples 8-3, $F = 83$ and the resultant diameter would be 0.375 m. In these examples, the results would indicate that a packed tower would be chosen for the duty on account of the resulting diameter. Full details of the design of packed and plate columns are available in many standard textbooks.

8.4.5 Height of column

In the case of a plate column, the height is simply the product of the number of plates required for the separation and the plate spacing. For a packed column, data are available for a number of systems providing the height equivalent to a theoretical plate ($HETP$). If the number of theoretical plates N is determined by a graphical method, the height of the packed section is then given by $Z = N(HETP)$. Graphical methods for determining the number of theoretical plates are included in chapter 7.

Where $HETP$ data are unavailable, use is made of the mass transfer rate expressions discussed in chapter 5 and the concepts of transfer unit are introduced. If the height of an overall transfer unit based on the gas phase is H_{OG} and the number of overall transfer units based on the gas phase is N_{OG} then the height of the packed section is given by

$$Z = N_{OG} \cdot H_{OG} \tag{8-11}$$

or, in terms of the liquid phase

$$Z = N_{OL} \cdot H_{OL} \tag{8-12}$$

The number of transfer units is defined by

$$N_{OG} = Z/H_{OG} = \int_{y_2}^{y_1} \frac{(1-y)_{lm}\,dy}{(y-y_e)(1-y)} \tag{8-13}$$

and

$$N_{OL} = Z/H_{OL} = \int_{x_2}^{x_1} \frac{(1-x)_{lm}\,dx}{(x_e - x)(1-x)} \tag{8-14}$$

where y is the mole fraction in the gas, x is the mole fraction in the liquid y_e the gas composition in equilibrium with liquid x, x_e the liquid composition in equilibrium with gas y, $(1-y)_{lm}$ the logarithmic mean of $(1-y)$ and $(1-y_e)$, and $(1-x)_{lm}$ the logarithmic mean of $(1-x)$ and $(1-x_e)$.

For dilute conditions where $(1-y)_{lm}/(1-y) \simeq 1$, eqn. 8-13 becomes

$$N_{OG} = \int_{y_2}^{y_1} dy/(y - y_e) \tag{8-15}$$

Equation 8-15 is frequently used in gas absorption problems but any of these equations may be evaluated by graphical integration to give the number of transfer units.

The height of a transfer unit H_{OG} or H_{OL} is defined by eqns. 8-16 and 8-17.

$$H_{OG} = G_m/K_G a \cdot P \tag{8-16}$$

$$H_{OL} = L_m/K_L a \cdot \rho_L \tag{8-17}$$

and in terms of the individual gas and liquid phases,

$$H_G = G_m/k_G a \cdot P \tag{8-18}$$

$$H_L = L_m/k_L a \cdot \rho_L \tag{8-19}$$

Equations 8-16 to 8-19 may be combined to give

$$H_{OG} = H_G + \frac{mGm}{L_m} \cdot H_L \tag{8-20}$$

$$H_{OL} = H_L + \frac{L_m}{mG_m} \cdot H_G \tag{8-21}$$

Experimental data are available for a number of systems to enable the required values of H_{OG}, H_{OL}, H_G and H_L to be used with confidence. In the absence of mass transfer data to enable the values to be calculated, empirical correlations may be consulted which relate the heights of individual phase transfer units to the gas and liquid flow rates, the type and size of packing, gas and liquid physical properties, and the prevailing conditions of temperature and pressure.

The following examples illustrate the principles involved in calculating packing heights by the method of transfer units.

8.4.6 Design examples

Example 8-5: A soluble gas is to be absorbed in a counter-current packed tower using a solute free liquid. The equilibrium relationship is given by $y_e = mx$. Obtain an expression for the number of transfer units if the inlet and exit gas compositions are y_1 and y_2 respectively. If a recovery of 90% of the solute is required and the liquid rate is 1.5 times the minimum calculate the number of transfer units required and compare this with the general graphical method.

Solution With reference to Fig. 8-7, a mass balance over the top section of the column gives

$$G_m(y - y_2) = L_m(x - x_2)$$

If inlet solvent is solute free, $x_2 = 0$ and

$$x = (G_m/L_m)(y - y_2)$$

The operating line equation is

$$y = (L_m/G_m)x + y_2$$

The equilibrium line equation is

$$y_e = mx = (mG_m/L_m)y - (mG_m/L_m)y_2$$

By definition the number of transfer units N_{OG} is given by

$$N_{OG} = \int_{y_2}^{y_1} dy/(y - y_e) \tag{8-15}$$

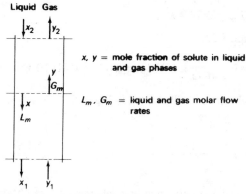

Liquid Gas

x, y = mole fraction of solute in liquid and gas phases

L_m, G_m = liquid and gas molar flow rates

Fig. 8-7

223

Hence

$$N_{OG} = \int_{y_2}^{y_1} \frac{dy}{y - \dfrac{mG_m}{L_m} \cdot y + \dfrac{mG_m}{L_m} \cdot y_2}$$

$$= \int_{y_2}^{y_1} \frac{dy}{\left(1 - \dfrac{mG_m}{L_m}\right) y + \dfrac{mG_m}{L_m} \cdot y_2} \qquad (8\text{-}22)$$

Integrating and writing $\lambda = mG_m/L_m$ gives

$$N_{OG} = \frac{1}{1 - \lambda} \ln\left[(1 - \lambda)\frac{y_1}{y_2} + \lambda\right] \qquad (8\text{-}23)$$

In the example: $y_1/y_2 = 10$, $L_m/G_m = 1 \cdot 5 (L_m/G_m)_{min}$, $x_2 = 0$

Now

$$(L_m/G_m)_{min} = (y_1 - y_2)/(x_{e1} - x_2)$$

where x_{e1} is the liquid composition in equilibrium with y_1 then

$$(L_m/G_m)_{min} = (y_1 - y_2)/(y_1/m) = [y_1 - (y_1/10)]/(y_1/m) = 0 \cdot 9\, m$$

and

$$\frac{L_m}{G_m} = 1 \cdot 5 (L_m/G_m)_{min} = 1 \cdot 5 \times 0 \cdot 9\, m = 1 \cdot 35\, m$$

then

$$mG_m/L_m = 0 \cdot 74 = \lambda$$

Hence the eqn. 8-23

$$N_{OG} = \frac{1}{1 - 0 \cdot 74} \times 2 \cdot 303 \log_{10}[(1 - 0 \cdot 74)\, 10 + 0 \cdot 74] = 4 \cdot 63$$

The graphical method is illustrated in Fig. 8-8.

As

$$N_{OG} = \int_{y_2}^{y_1} \frac{dy}{(y - y_e)},$$

values of y and the corresponding values of y_e are obtained from the graph. The area under a plot of $1/(y - y_e)$ against y between the values

224

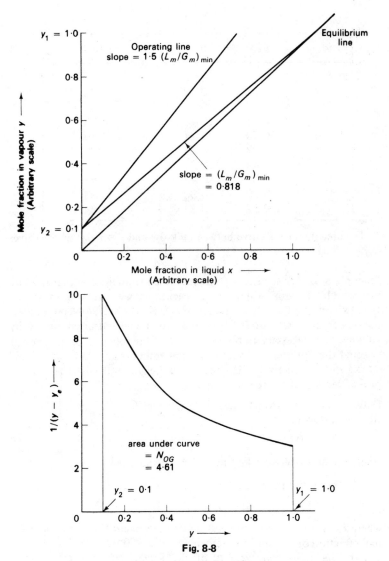

Fig. 8-8

of y_1 and y_2, then gives the value of N_{OG}. This method is applicable to all systems, but the expression for N_{OG} developed in this example only applies where both the equilibrium and operating lines are straight.

From Fig. 8-8, a table can be constructed with the values of y having an arbitrary scale.

y	y_e	$(y - y_e)$	$1/(y - y_e)$
0·1	0	0·10	10·0
0·2	0·08	0·12	8·33
0·3	0·15	0·15	6·67
0·4	0·21	0·19	5·26
0·5	0·29	0·21	4·76
0·6	0·37	0·23	4·35
0·7	0·44	0·26	3·85
0·8	0·52	0·28	3·57
0·9	0·60	0·30	3·33
1·0	0·67	0·33	3·03

The area under the curve between $y = 0·1$ and $y = 1·0$ (Fig. 8-8) gives $N_{OG} = 4·61$.

Example 8-6:   A solvent is to be recovered from a solvent-air mixture by scrubbing with water in a packed tower at a pressure of $101·3 \text{ kN/m}^2$ and at a temperature of 300 K. The solvent vapour enters the tower at a rate of 0·1 kg/s and has a concentration of 2% by volume. It is necessary to recover 99·9% of the solvent; for the packing selected the optimum gas and water flow rates are known to be 1·3 and $2·0 \text{ kg/sm}^2$ respectively. Calculate the height and diameter of the packed section of the tower required.

Data: (1)   The overall mass transfer coefficient $K_G \cdot a$ is related to the liquid flow rate by

$$K_G a = 0·0275 \, L^{0·5}$$

where the unit of $K_G a$ is kg/s m^3 (kN/m^2) and L is kg/s m^2.

(2)   The equilibrium data for the process are given by

$$P_e = 0·02 \, x$$

where P_e is the equilibrium pressure of the solvent (kN/m^2) and x the mole fraction of solvent in water where $0 < x < 0·03$.

(3)   The molecular weights of the solvent and air are 70 and 29 kg/kmol respectively.

Solution   The system is shown in Fig. 8-9.

It is necessary first of all to calculate the air flow rate from which the diameter may be obtained.

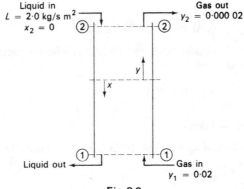

Liquid in
$L = 2\cdot0$ kg/s m^2
$x_2 = 0$

Gas out
$y_2 = 0\cdot000\ 02$

Liquid out

Gas in
$y_1 = 0\cdot02$

Fig. 8-9

At the gas inlet, $y_1 = 0\cdot02$ so that if the flow rate of air is a kg/s we can write

$$0\cdot02 = (0\cdot1/70)/[(0\cdot1/70) + (a/29)]$$

from which

$$a = 2\cdot03 \text{ kg/s.}$$

Hence the cross-sectional area A of the packing is given by

$$A = 2\cdot03/1\cdot3 = 1\cdot56 \text{ m}^2$$

and the diameter is $1\cdot41$ m

The height of packing may be obtained from the product of the height of a transfer unit and the number of transfer units,

$$Z = H_{OG} \cdot N_{OG} \tag{8-11}$$

The height of a transfer unit, H_{OG}, is defined by

$$H_{OG} = G/K_G a \cdot P \tag{8-16}$$

where G is the gas flow rate (kg/s m^2), $K_G a$ the overall mass transfer coefficient (kg/s m^3(kN/m^2)), and P the column pressure (kN/m^2).

In this example,

$$H_{OG} = 1\cdot3/101\cdot3\cdot K_G a$$

Now

$$K_G a = 0\cdot0275\ L^{0\cdot5} = 0\cdot0275 \times 2\cdot0^{0\cdot5} = 0\cdot0389 \text{ kg/s m}^2(\text{kN/m}^2)$$

and

$$H_{OG} = 1 \cdot 3/(0 \cdot 0389 \times 101 \cdot 3) = 0 \cdot 33 \text{ m}$$

The number of transfer units in conditions where concentrations are dilute may be defined as

$$N_{OG} = \int_{y_1}^{y_2} dY/(Y - Y_e) \qquad (8\text{-}15)$$

where Y is the mol ratio.

Now $P_e = 0 \cdot 02x$ and $y_e = P_e/P = 0 \cdot 02x/101 \cdot 3 = 0 \cdot 000197x$

Now $Y_e = \dfrac{Y_e}{1+Y_e}$ and $x = \dfrac{x}{1+x}$

and $\dfrac{Y_e}{1+Y_e} = \dfrac{0 \cdot 000197x}{1+x}$

$$y_1 = 0 \cdot 02, \text{ hence } Y_1 = 0 \cdot 02/(1 - 0 \cdot 02) = 0 \cdot 02041$$

$$y_2 = 0 \cdot 00002 \text{ and hence } Y_2 \approx 0 \cdot 00002$$

Taking a mass balance over the top part of the packed section shown in Fig. 8-9 gives

$$G_m(Y - Y_2) = L_m(X - X_2)$$

Now $X_2 = 0$ for solute free liquor

G_m = molar flowrate of air = $1 \cdot 3/29 = 0 \cdot 0448$ kmol/s m^2

L_m = molar flowrate of water = $2 \cdot 0/18 = 0 \cdot 111$ kmol/s m^2

Therefore

$$0 \cdot 0448(Y - 0 \cdot 000 \ 02) = 0 \cdot 111(X - 0)$$

from which

$$X = 0 \cdot 4036Y - 0 \cdot 000 \ 008$$

As

$$Y_e = 0 \cdot 000 \ 197X$$
$$Y - Y_e = Y - 0 \cdot 000 \ 197(0 \cdot 4036Y - 0 \cdot 000 \ 008)$$
$$= 0 \cdot 999921Y + 1 \cdot 5 \times 10^{-9}$$

From eqn. 8-15

$$N_{OG} = \int_{Y_1}^{Y_2} \frac{dY}{(0 \cdot 999921Y + 1 \cdot 5 \times 10^{-9})}$$

$$= \frac{1}{0 \cdot 999921} \ln \left[\frac{(0 \cdot 000021 \times 0 \cdot 02041 + 1 \cdot 5 \times 10^{-9})}{(0 \cdot 999921 \times 0 \cdot 000002 + 1 \cdot 5 \times 10^{-9})} \right]$$

$$= 6 \cdot 93$$

Hence the height of packing

$$Z = 6 \cdot 93 \times 0 \cdot 33 = 2 \cdot 29 \text{ m}$$

8.4.7 Non-isothermal absorption and heat balances

When dealing with dilute gas mixtures being absorbed in a liquid, the simplifying assumption of isothermal operation is frequently made. Whilst this assumption is frequently justified and valid, there are many situations where large heat effects are involved. A particular absorption duty, for example may be known to be highly exothermic or the absorption of large quantities of solute gas may give rise to concentrated solutions. If, as a result of the absorption, the liquid temperature rises, the equilibrium solubility of the solute will be appreciably reduced as will the absorber capacity. There are two ways of dealing with this situation and these are discussed briefly in the following section and will be illustrated by means of an example.

In the case of very high heat loads, the liquid may be removed at intervals from the tower, cooled externally, and returned to the tower. In this way, the operation of the tower may be considered isothermal. With reference to Fig. 8-10, a heat balance over the top section of the

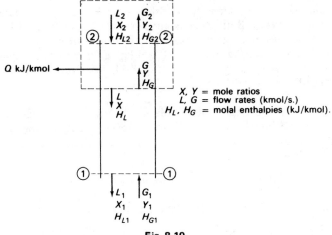

Fig. 8-10

Problems in heat and mass transfer

tower allows the heat load to be calculated, that is

$$H_{L2} \cdot L_2 + H_G \cdot G = H_{G2} \cdot G_2 + H_L \cdot L + Q \qquad (8\text{-}24)$$

where H_L, H_G are the molal enthalpies of the liquid and gas streams (kJ/kmol).

The molal enthalpy of the solution of composition X and temperature t_L may be calculated from

$$H_L = C_L (t_L - t_0) M_{av} + \Delta H_s \qquad (8\text{-}26)$$

where c_L is the specific heat of liquid (kJ/kg K), t_L, t_0 the solution and datum temperature of liquid (K), M_{av} the mean molecular weight of solution (kg/kmol), and ΔH_s = heat of mixing (kJ/kmol).

ΔH_s is equal to zero for ideal solutions and heat evolution in such cases is due to the latent heat of condensation of the absorbed solute alone.

If instead of isothermal operation where Q is removed, adiabatic operation ($Q = 0$) is to be preferred, we can use the above equations to calculate the temperature of the liquid L. It is necessary to know the temperature of the gas G as it enters the section which will in turn necessitate consideration of simultaneous heat, mass, and enthalpy balances. To simplify the calculation, it is often permissible to assume that the sensible heat change of the gas stream is unimportant and that all the heat evolution goes to raise the liquid temperature.

An example will now be considered to show how the construction of an adiabatic operation equilibrium curve differs from that applicable to isothermal operation.

Example 8-7: Moist air is to be dried using a 50% aqueous sodium hydroxide solution. Both streams enter a packed tower at 293 K and atmospheric pressure. The moisture content of the air is to be reduced from 0.015 kg water/kg dry air to 0.001 kg/kg. Properties of aqueous sodium hydroxide are presented in Fig. 8-11. Construct the adiabatic equilibrium curve for this operation.

Solution: The integral heat of solution is referred to solid sodium hydroxide and liquid water. The datum temperature t_0 will be chosen as 293 K and the solvent in the liquid phase is sodium hydroxide and the solute is water.

Basis of calculation: flow rate of solvent $L_s = 1.0$ kmol/s, and datum temperature $t_0 = 293$ K.

Rewriting eqn. 8-24 and putting $Q = 0$ for adiabatic operation,

$$H_L \cdot L - H_{L2} \cdot L_2 = H_G \cdot G - H_{G2} \cdot G_2 = G_s(H_G' - H_{G2}')$$

230

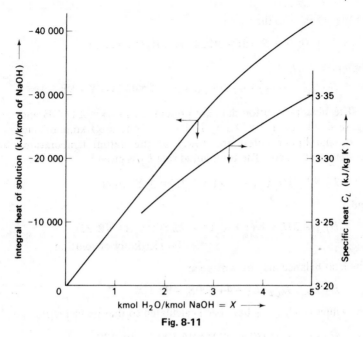

Fig. 8-11

where G_s is the rate of solvent gas (kmol/s), and H_G' the enthalpy of gas (kJ/kmol of dry air).

As the gas may be assumed to be at the base temperature $t_0 = 293$ K its enthalpy includes only the latent heat of vaporization at 293 K which is equal to 44 200 kJ/kg. Thus the enthalpy of the gas stream containing Y kmol water/kmol dry air $= 44\ 200\ Y = H_G'$.

So that

$$G_s(H_G' - H_{G2}') = G_s \cdot 44\ 200\ (Y - Y_2)$$

A mass balance on the water gives

$$G_s(Y - Y_2) = L_s(X - X_2) = 1 \cdot 0(X - X_2)$$

and hence

$$G_s(H_G' - H_{G2}') = (X - X_2)\ 44\ 200.$$

At the top of the tower

$$Y_2 = 0 \cdot 001(29/18) = 0 \cdot 0016 \text{ kmol } H_2O/\text{kmol air}$$
$$X_2 = (50/50)(40/18) = 2 \cdot 22 \text{ kmol } H_2O/\text{kmol NaOH}$$

231

At the gas inlet to the tower

$$Y_1 = 0.015(29/18) = 0.024 \text{ kmol } H_2O/\text{kmol air}$$

therefore

$$G_s(H_G' - H_{G2}') = 44\,200(X - 2.22) = 44\,200\,X - 98\,120$$

The heat of solution data are presented as kJ/kmol NaOH and this figure should be multiplied by $1/(1 + X)$ to obtain kJ/kmol of solution.

As the liquor enters the tower at the datum temperature, the sensible heat is zero. The total liquid rate L_2 is given by

$$L_2 = L_s(1 + X_2) = 1.0(1 + 2.22) = 3.22 \text{ kmol/s.}$$

and

$$H_{L2} = \Delta H_s \times 1/(1 + X_2) = -22\,800 \times 1(1 + 2.22)$$
$$= -7080 \text{ kJ/kmol of solution}$$

The heat balance may be written as

$$H_2 \cdot L - H_{L2} \cdot L_2 = 44\,200X - 98\,120$$

The values of H_{L2}, L_2 have been calculated so that we can write

$$H_L \cdot L - (-7080 \times 3.22) = 44\,200X - 98\,120$$

This equation relates the top conditions to those existing at a lower level in the tower where X will be greater than X_2 and the liquid temperature t_L is greater than the inlet temperature $t_o = 293$ K. A value of X will be assumed and values of L, H_s and H_L will be calculated for a new temperature t_L.

Let $X = 2.3$ kmol $H_2O/$kmol NaOH, then

$$L = L_s(1 + X) = 1.0(1 + 2.3) = 3.3 \text{ kmols solution}$$

and

ΔH_s from the data provided at $X = 2.3 = -22\,600$ kJ/kmol NaOH

ΔH_s for the solution $= -22\,600 \times 1/(1 + 2.3) = -6850$ kJ/kmol soln.

Now

$$H_L = C_L(t_L - t_0)M_{av} + H_s \text{ kJ/kmol} \qquad (8\text{-}26)$$

and

$$M_{av} = [(2.3 \times 18) + (1.0 \times 40)]/3.3 = 24.67 \text{ kg/kmol}$$

232

and

C_L from the data provided = 3·282 kJ/kg K

Hence

$$H_L = 3·282(t_L - 293)24·67 - 6850 = 81·0\ t_L - 30\ 600.$$

Substituting H_L and L in the heat balance of eqn. 8-24 gives

$$(81·0t_L - 30\ 600)\ 3·3 - (-7080 \times 3·22) = 44\ 200 \times 2·3 - 98\ 120$$

from which

$$t_L = 305·7 \text{ K}$$

To locate the top condition and the point just calculated on an X-Y diagram, the equilibrium partial pressure of water is estimated from data such as are shown in Fig. 8-12 to give at $X = 2·3$, $t_L = 305·7$ K, $P_e = 0·35$ kN/m^2.

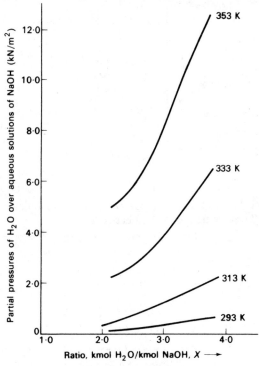

Fig. 8-12

Hence

$$Y^* = 0.35/(101.3 - 0.35) = 0.00\ 347\ \text{kmol H}_2\text{O/kmol air.}$$

At $X = 2.22$, $t = 293$ K, $P_e = 0.12$, and

$$Y^* = 0.12/(101.3 - 0.12) = 0.001\ 19$$

These points can be located on the required $X - Y$ diagram.
The condition of the tower as calculated and shown in Fig. 8-13.
The conditions at level a have just been calculated and in a similar way the values at level b will now be calculated.

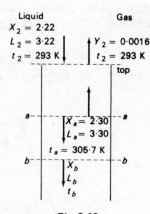

Fig. 8-13

The heat balances between levels a and b (assuming that the gas remains at the same temperature) become

$$H_{Lb} \cdot L_b - H_{La} \cdot L_a = 44\ 200\ X - 98\ 120$$

Choose $X = 2.35$

then

$$L_b = (L_s(1 + X) = 1.0(1 + 2.35) = 3.35\ \text{kg/s}$$

At $X = 2.35$ $\Delta H_{sb} = -23\ 000$ kJ/kmol NaOH

$$= -23\ 000/(1 + 2.35)$$

$$= -6870\ \text{kJ/kmol of solution}$$

also

$$M_{avb} = [(2 \cdot 35 \times 18) + (1 \times 40)] / 3 \cdot 35 = 24 \cdot 57$$
$$C_{Lb} = 3 \cdot 284 \text{ kJ/kg K}$$

and

$$H_{Lb} = 3 \cdot 284(t_b - 293)24 \cdot 57 - 6870 = (80 \cdot 69 t_b - 30\,510)$$

Hence

$$(80 \cdot 69 t_b - 30\,510)3 \cdot 35 - (81 \cdot 0 \times 305 \cdot 7 - 30\,600)3 \cdot 30$$
$$= 44\,200 \times 2 \cdot 35 - 98\,120$$

from which

$$t_b = 328 \cdot 1 \text{ K}$$

At $X = 2 \cdot 35$, $t_b = 328 \cdot 1$, $P_e = 2 \cdot 0 \text{ kN/m}^2$, hence

$$Y_b^* = 2 \cdot 0/(101 \cdot 3 - 2 \cdot 0) = 0 \cdot 020$$

The adiabatic equilibrium curve is plotted in Fig. 8-14 and for comparison the isothermal equilibrium curve at 293 K may be calculated.

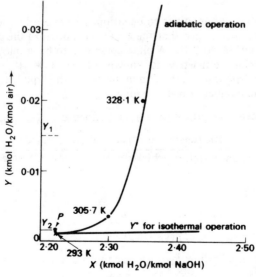

Fig. 8-14

At $X = 2\cdot4$, $P_e = 0\cdot15$ kN/m^2, hence

$Y^* = 0\cdot001\ 48$

The equilibrium line is thus drawn in, and bearing in mind that the operating line will be drawn from point P (at the top of the tower) to the ordinate at Y_1; we can now see the startling effect of adiabatic operation.

8.5 Column efficiencies

When a calculation has involved the estimation of the number of theoretical plates, the assumption has been made that liquid and vapour attain a state of equilibrium. In practice, this state is not reached and in an actual column the number of plates required to effect the given duty will always be greater than the number of theoretical plates calculated. Thus, the concept of efficiency is introduced. If the number of theoretical plates is divided by the number of actual plates, the overall column efficiency is obtained, thus averaging the conditions occurring on each individual plate. The efficiency of one plate in a column is a measure of the proximity to equilibrium conditions which is attained on that plate. This efficiency is known as the Murphree plate efficiency and its definition and application is shown in the following example.

Example 8-8: Ammonia is to be stripped from a dilute ammonia/air mixture with water such that the ammonia content of the gas stream is reduced from 3% to 0.1%. A plate column is to be employed and the Murphree plate efficiency is known to be 35%. If there is no appreciable vapour pressure of ammonia above the liquid at equilibrium, calculate the number of plates required.

The Murphree plate efficiency E_{MV} is defined as

$$\frac{100\ (\text{change in concentration achieved})}{\text{change in concentration achieved if the system reached equilibrium}}$$

(8-27)

Solution: From Fig. 8-15

$E_{MV} = (y_n - y_{n+1})/(y_n - y_e)$

However, at equilibrium, $P_{NH_3} = 0$ and $y_e = 0$, then

$0\cdot35 = (y_n - y_{n+1})/y_n$

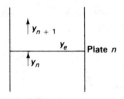

Fig. 8-15

from which

$$y_{n+1} = 0{\cdot}65\, y_n$$

At the bottom of the tower, $y_n = y_1 = 0{\cdot}03$ so that the vapour leaving plate 1 is

$$y_{n+1} = 0{\cdot}65 \times 0{\cdot}03 = 0{\cdot}0195$$

and, in a similar way, the table below can be constructed.

The vapour leaving plate 8 contains 0.096% ammonia which meets the specification required and hence 8 plates are necessary.

By its definition, the Murphree plate efficiency is not the same as the overall column efficiency, and the difference between the two is illustrated below in the graphical construction of Example 8-9.

Plate Number (n)	Vapour rising to plate n y_n	Vapour leaving plate n $y_{n+1} = 0{\cdot}65\, y_n$
1	0·03	0.0195
2	0·0195	0·0127
3	0·0127	0·008 26
4	0·008 26	0·005 36
5	0·005 36	0·003 48
6	0·003 48	0·002 26
7	0·002 26	0·001 47
8	0·001 47	0·000 96

Example 8-9: A solvent is to be recovered from a gas stream by absorption with clean water in a plate column. The solvent enters at a concentration of 2% by volume and the maximum allowable loss is 0.1%. A liquid rate of 1.3 times the minimum rate is employed and the Murphree efficiency is known to be 50% based on the gas phase.

Calculate the minimum number of plates to effect the duty. What is the overall column efficiency under these conditions? The equilibrium data are represented by

$$y = 0.5 \, x^2$$

where y, x are mole fractions of the solvent in the vapour and liquid phases respectively.

Solution: The Murphree plate efficiency based on the gas phase may be defined as

$$E_{MV} = 100(y_n - y_{n-1})/(y_n^* - y_{n-1}) \qquad (8\text{-}28)$$

where y_n^* is the composition of the hypothetical vapour phase that would be in equilibrium with the liquid composition leaving the actual stage. In a similar way, the Murphree efficiency based on the liquid phase, E_{ML} can be defined. In terms of the graphical method for stepping off stages, the efficiency dictates the percentage of the distance from the operating line to the equilibrium line to be taken as shown in Fig. 8-16. Thus a new 'equilibrium' line can be drawn if the Murphree efficiency is known and the actual number of plates obtained by graphical means.

The equilibrium data are plotted in Fig. 8-17 as the curve OA. The value of $(L/G)_{min}$ is obtained from the slope of the line joining the point (0,0.001) and the intersection of the equilibrium line and the line $y = 0.02$, i.e. the line OA.

This slope is found to be 0.095.

Using $1.3 \times (L/G)_{min}$ gives an operating line of slope equal to 0.1235 and this is drawn on Fig. 8-17 as BC. The efficiency based on

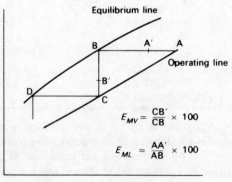

$$E_{MV} = \frac{CB'}{CB} \times 100$$

$$E_{ML} = \frac{AA'}{AB} \times 100$$

Fig. 8-16

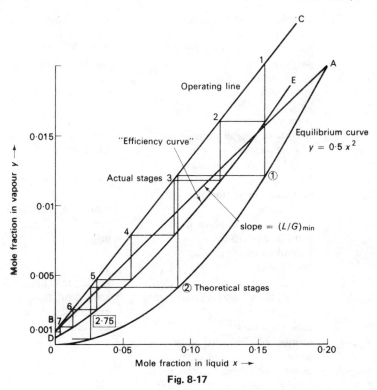

Fig. 8-17

the gas phase is 50% so that a series of points is obtained mid-way between the equilibrium and operating lines. This is shown as line DE on Fig. 8-17. The number of plates is obtained by stepping off between the operating line and the equilibrium line in the normal way for the two cases of 100% efficiency and $E_{MV} = 50\%$. It is found graphically that the number of theoretical plates is 2·75 and the number of actual plates is 7·3.

Hence the overall column efficiency is

$$100 \times 2 \cdot 75 / 7 \cdot 3 = 37 \cdot 7\%$$

List of symbols

a	specific area of packing	m^2/m^3
C_L	specific heat of liquid	kJ/kg K

E_{MV}, E_{ML}	Murphree plate efficiency	–
F	packing factor – Fig. 8-5	–
g	acceleration due to gravity	m/s^2
G	gas flow rate	$kmol/s$, $kmol/sm^2$
G'	gas flow rate – Fig. 8.5	$kg/s\,m^2$
H,H'	enthalpies	$kJ/kmol$, kJ/kg
H_G, H_L	heights of individual phase transfer units	m
H_{OG}, H_{OL}	heights of overall transfer units	mm
ΔH_s	heat of mixing	$kJ/kmol$
k_G	individual gas phase mass transfer coefficient	$kmol/s\,m^2\,(kN/m^2)$
$k_G a$	individual mass transfer coefficient	$kmol/s\,m^3\,(kN/m^2)$
k_L	individual liquid phase mass transfer coefficient	m/s
$k_L a$	individual mass transfer coefficient	$1/s$
$K_G a, K_L a$	overall mass transfer coefficients	$kmol/s\,m^3\,(kN/m^3)$, l/s
K_G, K_L	overall mass transfer coefficients	$kmol/s\,m^2\,(kN/m^2)$, m/s
L	liquid flow rate	$kmol/s$, $kmol/s\,m^2$
M	molecular mass	$kg/kmol$
N_G, N_L	number of individual phase transfer units	–
N_{OG}, N_{OL}	number of overall transfer units	–
P	total pressure, partial pressure	kN/m^2
P_e	equilibrium partial pressure	kN/m^2
P^0	vapour pressure of pure component	kN/m^2
Q	heat evolved or removed	$kJ/kmol$, kJ/kg
t	temperature	K
x	liquid phase composition-mole fraction	–
x_e	equilibrium composition	–
X	mole ratio	–
y	vapour phase composition-mole fraction	–
y_e	equilibrium composition	–
y	mole ratio	–
Z	height of packed section	m
ρ	density	kg/m^3
μ	viscosity	$N\,s/m^2$

ψ ratio density of water/density
of liquid in Fig. 8-5   —

λ ratio mG_m/L_m —

Subscripts

1, 2   components or conditions 1 and 2

m molal

min   minimum

s solute

Further Reading

BACKHURST, J. R. & HARKER, J. H., *"Process Plant Design"* Heinemann Educational Books London. 1973

BADGER, W. L. & BANCHERO, J. T. *"Introduction to Chemical Engineering"* pp. 418–469, McGraw-Hill Book Co. New York 1955.

COULSON, J. M., RICHARDSON, J. F., BACKHURST, J. R. & HARKER, J. H. *"Chemical Engineering"* Vol. 2. 3rd Ed. Pergamon Press, Oxford 1979.

LEVA, M, *"Tower Packings and Packed Tower Design"* US. Stoneware Co. Ohio U.S.A. 1953.

MORNS, G. A. & JACKSON, J. *"Absorption Towers"* Butterworths Scientific Publications London 1953.

NORMAN, W. S. *"Absorption, Distillation and Cooling Towers"* Longmans London 1961.

PERRY, J. H. Ed. *"Chemical Engineers Handbook"* 4th Ed. pp. 14–24 – 14–40 McGraw-Hill Book Co. New York 1963.

SHERWOOD, T. K. & PIGFORD, R. L. *"Absorption and Extraction"* 2nd Ed. McGraw-Hill Book Co. New York. 1952.

SAWISTOWSKI, H. & SMITH, W. *"Mass Transfer Process Calculations"* Interscience New York and London 1963.

TREYBAL, R. E. *"Mass Transfer Operations"* 2nd Ed. McGraw-Hill Book Co. New York 1963.

Problem 8-1: A soluble gas is to be absorbed from air in water in a packed tower. The water enters the tower solute free and leaves with a concentration of $0·08$ kmol solute/kmol of water. The inlet and exit gas stream concentrations are $0·08$ and $0·009$ kmol solvent/kmol of air. The heights of the individual gas and liquid phase transfer units are $1·0$ and $0·5$ m respectively and the equilibrium relationship is given by $Y_e = 0·06X$. Calculate the packed height in the tower.

$(4·7 \text{ m})$

Problem 8-2: A soluble gas is to be absorbed from an air gas mixture in a countercurrent packed tower using solute free liquid. The equilibrium relationship is given by $y_e = mx$. Show that the number of overall gas transfer units N_{OG} is given by

$$N_{OG} = [1/(1 - \lambda)] \ln [(1 - \lambda)(y_1/y_2) + \lambda]$$

If 99% of the solute is to be recovered using a liquid rate 1.75 times the minimum, and the height of a transfer unit is 1.0 m, what packed height is required?

(8.9 m)

Problem 8-3: An air stream containing 4% (by volume) SO_2 and 3% water is to be scrubbed with water in a packed tower to give a solution of 0.5% (by weight) of SO_2. The gas flow rate is 0.826 m^3/s and the temperature is 298 K. The exit concentration of SO_2 is to be 0.5% and the density of SO_2 is 2.68 kg/m^3. The following packing data are available

Specific area of 50 mm rings = 95.2 m^2/m^3

Gas/Liquid ratio (V_g/V_l) at the loading point	35	45	60
Wetting rate (m^3/s) m x 10^4	2·19	1·81	1·44

Determine the rate of absorption and the cross sectional area of the packing.

(0.078 kg/s; 1.5 m^2)

Problem 8-4: Benzene is to be stripped from a benzene-air mixture in a packed tower at a total pressure of 101.3 kN/m^2 using a non-volatile oil (molecular weight = 200) as the absorbent. The inlet concentration of benzene is 8.0 vol % and 92% is to be recovered. The total gas feed to the tower is 0.19 kg/s and the inlet oil contains 0.5 mol % of benzene. The vapour pressure of benzene at the tower temperature is 13.3 kN/m^2. Estimate the minimum oil flow rate to achieve this duty and the composition (mol %) of the exit oil stream at this condition.

(0.055 kg/s; 60.8%)

Problem 8-5: Using the data of Example 8-7, construct the adiabatic equilibrium curve for the absorption of water from moist air when the inlet and exit moisture contents are 0.02 and 0.0005 kg/kg respectively. 40% aqueous sodium hydroxide is to be used and both streams enter at 293 K.

9. Solvent Extraction

9.1 Liquid-liquid extraction

Solvent extraction, in which one component is recovered from a mixture of liquids by the addition of a solvent in which the component is soluble, is used where evaporation is too expensive or where the components of the mixture have approximately equal or low volatilities. In addition, solvent extraction is used for separating heat sensitive materials or where the solute is present in small concentrations. Important industrial applications of the operation include the recovery of acetic and lactic acids, the separation of fissile materials and the Edelanu process in petroleum refining.

9.1.1 Operations involved

Referring to Fig. 9-1, the separation of a product B from a feed stream A involves three stages. Firstly, mixing with a solvent S, with which in theory, A should be completely immiscible though this is never the case; secondly, the settling stage in which the extract phase (the exit solvent stream) is separated from the raffinate (the exit treated feed stream); and thirdly, solvent recovery and recycle. There are four ways in which these three operations may be carried out as shown in Fig. 9-2:

(a) *Single contact* In this system, the efficiency of mass transfer depends upon the equilibrium conditions, the degree of agitation in the mixer and the ratio f/s.

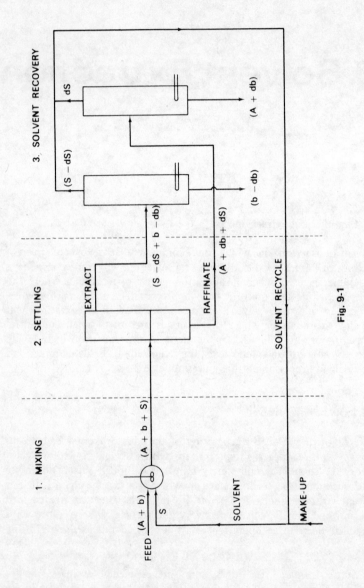

Fig. 9-1

244

(a) single contact

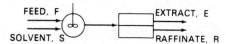

(b) multiple co-current contact

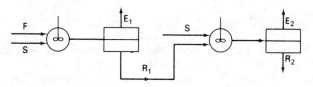

(c) multiple counter-current contact

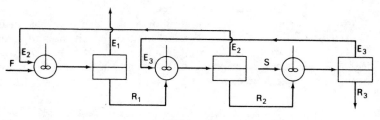

Fig. 9-2

(b) *Multiple co-current contact* This is a costly mode of operation with two stages being the most economical layout. The maximum extraction efficiency is obtained where equal amounts of solvent are introduced into each mixer.

(c) *Multiple counter-current contact* This is a highly efficient arrangement, usually involving 2 to 5 stages.

(d) *Continuous counter-current contact* The solvent is introduced at the bottom of a column as a dispersion of drops which are allowed to rise through the feed, which is the continuous phase fed at the top of the column. The column may be packed with ceramic rings for example, or some other device such as rotary discs, pulsed plates, or a rotary annulus may be used for continuously mixing the two phases and allowing separation to take place.

9.2 Calculation of the number of theoretical stages for a given separation

Solvent extraction is impossible in a completely homogeneous system and two cases are of practical importance.

9.2.1 Feed and solvent totally immiscible

(a) Co-current multiple contact

Example 9-1: 3·5 kg/s of a liquid feed containing a solute B dissolved in component A is to be treated with a solvent S in a four stage co-current contact unit. The concentration of B in the feed is 28·6% by weight. If 1·5 kg/s solvent S is added to each stage, what is the flowrate and composition of the raffinate leaving the fourth stage? Equilibrium data for the system are given in Table 9-1.

Solution The equilibrium data are plotted in Fig. 9-3 as mass of B /unit mass of S in extract (y) against mass of B/unit mass of A in raffinate (x). Referring to Fig. 9-2(b), if x_F = mass of B/mass of A in feed F, a = mass flow of component A, x_1 = mass of B/mass of A in raffinate, R s = mass flow of solvent S, and y_1 = mass of B/mass of S in

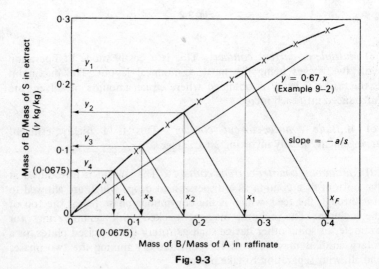

Fig. 9-3

Table 9-1. Equilibrium Data for Example 9-1

Mass of B in raffinate (x kg/kg A)	Mass of B in extract (y kg/kg S)
0	0
0·05	0·05
0·10	0·096
0·15	0·135
0·20	0·170
0·25	0·203
0·30	0·232
0·35	0·256
0·40	0·275
0·45	0·280

extract E, then a balance for component B gives

$$ax_F = ax_1 + sy_1 \qquad (9\text{-}1)$$

or

$$y_1/(x_1 - x_F) = -a/s$$

In this example, the concentration of B in the feed is 28·6%, therefore

$$x_F = 0·286/(1 - 0·286) = 0·4 \text{ kg/kg}$$

The flow of component A is given by

$$a = 3·5(1 - 0·286) = 2·5 \text{ kg/s}$$

and

$$s = 1·5 \text{ kg/s}$$

Therefore, for stage 1

$$-a/s = (-2·5/1·5) = -1·67.$$

A line of slope $-1·67$ is drawn in Fig. 9-3 through the point $(x_F = 0·4, y = 0)$ and the point where this intersects the eqiulibrium line gives the concentrations of B in the extract, $y_1 = 0·215$ kg/kg S, and in the raffinate, $x_1 = 0·274$ kg/kg A. It is now possible to proceed in the same way for the three remaining stages. As the amount of solvent S fed to each stage is the same, the slope $-a/s = -1·67$ in each case. The diagram is completed as shown giving

$$y_4 = 0·0675 \text{ kg/kg S}$$

and

$$x_4 = 0 \cdot 0675 \text{ kg/kg A}$$

The flow of A leaving stage 4 is the same as in the feed, that is
$a = 2 \cdot 5$ kg/s
and hence, in stream R_4, flow of B is

$$2 \cdot 5 \times 0 \cdot 0675 = 0 \cdot 17 \text{ kg/s.}$$

Thus the total flow of raffinate from stage 4 is

$$(2 \cdot 5 + 0 \cdot 17) = 2 \cdot 67 \text{ kg/s}$$

The concentration of B in the raffinate is

$$(0 \cdot 17 \times 100)/2 \cdot 67 = 6 \cdot 4\% \text{ by mass}$$

Example 9-2: If the equilibrium conditions for the system in Example 9-1 are represented by the equation $y = 0 \cdot 67 x$, how many theoretical stages would be required to achieve the same recovery of B, the flowrates remaining the same?

Solution   Where the equilibrium is linear and of slope m, a balance for component B across the first stage is now

$$ax_F = ax_1 + sy_1 = x_1(a + sm) \tag{9-1}$$

therefore

$$x_1 = ax_F/(a + sm) \tag{9-2}$$

For the second stage, assuming s remains the same

$$ax_1 = ax_2 + sy_2 = x_2(a + sm)$$

therefore

$$x_2 = ax_1/(a + sm)$$

or, substituting for x_1 from eqn. 9-2,

$$x_2 = x_F (a/(a + sm))^2 \tag{9-3}$$

Thus for n stages,

$$x_n = x_F (a/(a + sm))^n \tag{9-4}$$

and, solving for n

$$n = \log_{10}(x_n/x_F)/\ln(a/(a + sm)) \tag{9-5}$$

248

In this case $x_F = 0.4$ kg/kg A and $x_n = 0.0675$ kg/kg A, then $a = 2.5$ kg/s, $b = 1.5$ kg/s and hence in eqn. 9-5

$n = \ln(0.0675/0.4)/\ln(2.5/(2.5 + 1.5 \times 0.67)) = 5.15$ stages

Assuming an efficiency of say 80%

$(5.15/0.80) = 6.5$, say 7 stages would be specified.

(ii) Counter-current multiple contact

Example 9-3: An aqueous feed containing 28.6 kg of a solute B/100 kg solution is to be treated counter-currently with a solvent S in order to reduce the solute concentration to 9.1 kg/100 kg solution. A mixer-settler installation equivalent to five theoretical stages is available for this duty. If the solvent initially contains 4.75% solute by mass, what solvent aqueous phase flow ratio should be used and what is the estimated composition of the extract phase leaving the plant? The solvent and aqueous phases may be regarded as immiscible and the equilibrium data given in Table 9-1 apply.

Solution Using the nomenclature of the previous examples, where a kg/s is the flow of water in the aqueous phase, a mass balance over the first stage (Fig. 9-2(c)) gives

$f + e_2 = r_1 + e_1$

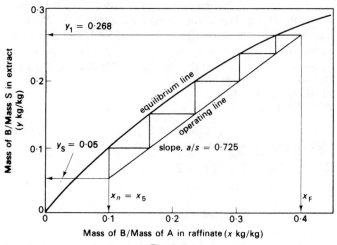

Fig. 9-4

or for component B,

$$ax_F + sy_2 = ax_1 + sy_1 \tag{9-6}$$

Similarly, for the entire unit

$$f + s = r_n + e_1$$

or for component B

$$ax_F + sy_s = ax_n + sy_1 \tag{9-7}$$

where y_S is the mass of B/mass of S in the solvent feed. Thus, from eqn. 9-7

$$y_S = (a/s)(x_n - x_F) + y_1 \tag{9-8}$$

This is a straight line of slope (a/s) known as the operating line.
In the present case,

$$x_F = 0.286/(1 - 0.286) = 0.4 \text{ kg/kg}$$
$$x_n = 0.091/(1 - 0.091) = 0.1 \text{ kg/kg}$$

and

$$y_S = 0.0475/(1 - 0.0475) = 0.05 \text{ kg/kg}$$

From eqn. 9-8, the operating line passes through the points (x_F, y_1) and (x_n, y_S). On Fig. 9-4, therefore, a line is drawn through the point $x_n = 0.1$, $y_S = 0.05$ such that it intersects with the line $x_F = 0.4$. The stages are then stepped off as shown. Adopting a trial and error approach, the slope of the operating line is varied until the number of theoretical stages is 5. This occurs when the slope $a/s = 0.725$, or
1 kg water should be associated with $1/0.725 = 1.38$ kg solvent,
1 kg water is associated with $(1 + 0.4) = 1.4$ g aqueous phase,
1.38 kg solvent is associated with $1.38(1 + 0.05) = 1.45$ kg solvent phase.
Therefore, solvent/feed ratio is
$1.45/1.4 = 1.035$

From Fig. 9-4, the mass of B/mass of S in the extract leaving the first stage is 0.268 kg/kg, and the concentration of B in the extract is

$$0.268 \times 100/(1 + 0.268) = 21.2\% \text{ by mass}$$

9.2.2 Feed and solvent partially miscible

In this case, it is convenient to use triangular diagrams to represent the equilibrium data and a typical example is shown in Fig. 9-5 for the

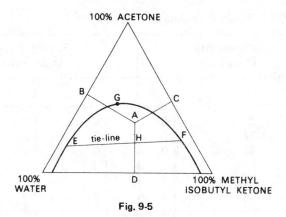

100% ACETONE

100% WATER

100% METHYL ISOBUTYL KETONE

Fig. 9-5

system acetone – water – methyl isobutyl ketone. Acetone, the solute, is completely miscible with the two solvents, water and MIK, but the two solvents are only partially miscible with each other. On this type of diagram, each corner of the triangle represents 100% of one component and hence a mixture by point A consists of acetone, water and MIK in the ratio of the lengths of the perpendiculars, AD, AC and AB. The curve shown is known as the binodal curve and the area below this represents a region in which a mixture will split up into two layers in equilibrium with each other. For example, a mixture of composition H will split up into two layers of compositions E and F and

$$\frac{\text{amount of layer of composition E}}{\text{amount of layer of composition F}} = \frac{\text{HF}}{\text{HE}}$$

The line EHF is known as a tie-line and the coordinates of such a line are found experimentally. The point G represents the one point on the binodal curve where a mixture does not split into two phases and this, also determined experimentally, is known as the plait point. As with a totally miscible feed and solvent, it is convenient to consider the problem according to the mode of operation; either co-current or counter-current.

(i) *Co-current multiple contact*

Example 9-4:   The tie-line data shown in Table 9-2 are available for a partially miscible system consisting of solvents B and C, and a solute A. 2·0 kg/s of a feed containing 60% A and 40% B is to be extracted by solvent C in a co-current contactor equivalent to three theoretical stages; the flow of solvent C being 0·91 kg/s to each stage. What are the

251

Table 9-2   Equilibrium Data for Example 9-4

Raffinate Phase (% by mass)			Extract Phase (% by mass)		
A	B	C	A	B	C
70	25	5·0	71·6	4·8	23·6
60	37	3·0	62·5	2·5	34·0
50	48	2·0	52·0	3·1	44·9
40	58·5	1·5	41·9	3·1	55·0
30	68·5	1·5	31·9	3·0	65·1
20	79	1·0	22·0	2·9	75·1
10	89	1·0	11·1	2·1	86·8

compositions of the raffinate and the extract leaving the third stage and what is the flowrate of each stream? What is the maximum possible concentration of A in the feed which could be handled?

Solution   The solution is carried out graphically as shown in Fig. 9-6. The point F represents the feed, 60% A and 40% B, and a line is drawn joining FC; point C representing the pure solvent. Flow of feed $f = 2·0$ kg/s, flow of solvent $s = 0·91$ kg/s, therefore

$$f/s = 2·0/0·91 = 2·2$$

and hence point M_1 is placed on FC such that $M_1C/M_1F = 2·2$. M_1 thus represents the mixture obtained when the feed and solvent are brought into contact in the first stage. M_1 lies on the tie line R_1E_1 which points represent the compositions of the raffinate and the extract leaving the first stage respectively.

By measurement from the figure

$$M_1E_1/M_1R_1 = (1·81/2·44) = 0·741$$

or

$$r_1/e_1 = 0·741$$

A mass balance over the first stage gives

$$r_1 + e_1 = f + s = (2·0 + 0·91) = 2·91 \text{ kg/s}$$

Therefore

$$0·741 \, e_1 + e_1 = 2·91$$

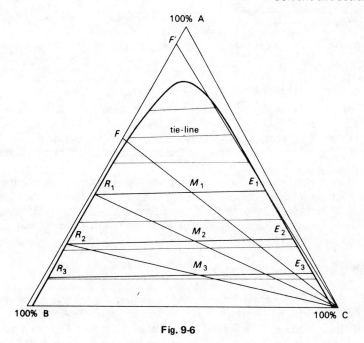

100% A

F′

F

tie-line

R_1 M_1 E_1

R_2 M_2 E_2

R_3 M_3 E_3

100% B   100% C

Fig. 9-6

and

$$e_1 = 1 \cdot 67 \text{ kg/s}, r_1 = 1 \cdot 24 \text{ kg/s}.$$

The line $R_1 C$ is now drawn in and point M_2 located by

$$M_2 C / M_2 R_1 = r_1 / s_2 = (1 \cdot 24 / 0 \cdot 91) = 1 \cdot 36$$

Proceeding as before, $e_2 = 1 \cdot 24 \text{ kg/s}, r_2 = 0 \cdot 91 \text{ kg/s}$ and

$$M_3 C / M_3 R_2 = r_2 / s_3 = (0 \cdot 91 / 0 \cdot 91) = 1 \cdot 0$$

Finally for the third stage,

$$M_3 E_3 / M_3 R_3 = (2 \cdot 88 / 3 \cdot 75) = 0 \cdot 768$$

or

$$r_3 / e_3 = 0 \cdot 768$$

As before,

$$r_3 + e_3 = r_2 + s_2 = (0 \cdot 91 + 0 \cdot 91) = 1 \cdot 82 \text{ kg/s}$$

therefore

$$0.768\, e_3 + e_3 = 1.82$$

then

$$e_3 = 1.03 \text{ kg/s and } r_3 = 0.79 \text{ kg/s}$$

Thus the flow of raffinate and extract from stage 3 are 0.79 kg/s and 1.03 kg/s respectively.

From Fig. 9-6, the composition of the raffinate, R_3 is 11% A, 87% B and 2% C and the composition of the extract, E_3 is 12% A, 3% B and 85% C.

The maximum concentration of A which can be handled in the feed is given by joining $E_1 C$ and projecting this to F'. The feed composition F' is equivalent to 96% A, 4% B. In dealing with such a feed, sufficient solvent C would have to be added to bring the mixture M into the two-phase region.

(ii) Counter-current multiple contact

Example 9-5: Repeat Example 9-4 assuming counter-current operation. The feed is as before, 2.0 kg/s of composition 60% A, 40% B, and the total amount of solvent C remains as before. If the maximum concentration of A in the raffinate leaving the plant is as for the co-current contactor, how many theoretical stages are required?

Solution Referring to Fig. 9-2(c), a mass balance over the first stages gives

$$f + e_2 = r_1 + e_1$$

or

$$f - e_1 = r_1 - e_2 = p \text{ (say)}$$

Similarly for the second stage,

$$r_1 + e_3 = r_2 + e_2$$

or

$$r_1 - e_2 = r_2 - e_3 = p$$

and for stage n,

$$r_{n-1} - e_n = r_n - e_{n+1} = p$$

In other words, the difference between the amount of raffinate leaving a stage and the amount of extract entering from the next stage is

254

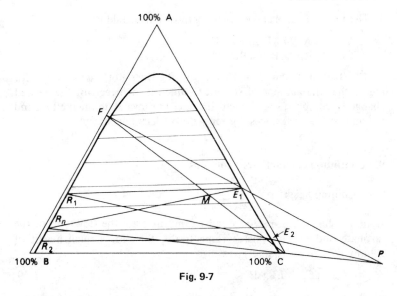

Fig. 9-7

constant and equals p which represents an imaginary mixture. Thus a line joining the points R_n and E_{n+1} will pass through a common pole P. For this example, the equilibrium data from Table 9-2 are plotted in Fig. 9-7. The feed is given as 60% A, 40% B and this is represented by point F and the line FC drawn in.

In Example 9-4, 0·91 kg/s solvent was fed to each of the three stages and hence total solvent flow = $(3 \times 0·91) = 2·73$ kg/s. Flow of feed $f = 2·0$ kg/s. Hence point M is obtained by dividing FC in the ratio

$$FM/MC = s/a = (2·73/2·0) = 1·37$$

The exit raffinate in Example 9-4 had the composition 11% A, 87% B and 2% C. This is represented by point R_n and indicates the desired extraction of A. R_nM is joined and projected to the binodal curve which it intersects at E_1. As in the previous example, R_1 is determined by drawing in the tie-line through E_1. The line FE_1 is then projected to meet the line R_nC projected at the pole, P. The extract composition from the second stage, E_2 is then found by the intersection of R_1P. The raffinate in equilibrium with E_2 is obtained by drawing in the tie-line E_2R_2.

R_2 represents a composition with a concentration of A less than the maximum which can be tolerated and hence two theoretical stages would be required.

255

The final raffinate and extract compositions would be

R_2: 5% A, 94% B and 1% C
E_2: 6% A, 2% B and 92% C.

The fact that the final tie-line, $E_2 R_2$ does not pass through R_n means that the amount of solvent C added is incorrect for the desired change in composition, though this does not invalidate the method and the number of ideal stages is given with sufficient accuracy.

9.3 Continuous extraction in columns

9.3.1 Column height

If in a column, L_R m³/m²s and L_E m³/m²s are the flowrates of the raffinate and extract phases respectively, then for a small height of column dZ a mass balance gives

$$L_R \cdot dC_R = L_E \cdot dC_E \tag{9-9}$$

where C_R and C_E kmol/m³ are the concentrations of solute in the raffinate and extract phases. As in absorption (chapter 8), the rate of transfer across the raffinate and extract films may be expressed in the form

$$N = k_R(C_R - C_{Ri}) = k_E(C_{Ei} - C_E) \text{ kmol/m}^2\text{s} \tag{9-10}$$

where k_R and k_E m/s are the mass transfer coefficients for the raffinate and extract films and the subscript i represents conditions at the interface.

Substituting from eqn. 9-10 into eqn. 9-9

$$L_R dC_R = k_R(C_R - C_{Ri})adZ \text{ kmol/m}^2 \text{ s}$$

or integrating between condition 1 at the bottom of the column and 2 at the top

$$\int_{C_{R_2}}^{C_{R_1}} dC_R/(C_R - C_{Ri}) = (k_R a/L_R)Z \tag{9-11}$$

The integral on the left hand side of eqn. 9-11 is the number of raffinate film transfer units, $(NTU)_R$ and $(L_R/k_R \cdot a) = (HTU)_R$, the height of a raffinate film transfer unit. By analogy with absorption, film and overall transfer units based on concentrations in both the raffinate and extract phases may be obtained and these are linked by

$$(HTU)_{OR} = (HTU)_R + (L_R/mL_E)(HTU)_E \qquad (9\text{-}12)$$

and

$$(HTU)_{OE} = (HTU)_E + (mL_E/L_R)(HTU)_R \qquad (9\text{-}13)$$

where m is the slope of the equilbrium line. The ratio (mL_E/L_R) is only constant with dilute solutions and L_R and L_E may be assumed constant only where a small amount of solute is transferred from one phase to the other. For dilute solutions and a linear equilibrium relationship

$$L_R(C_{R_1} - C_{R_2}) = K_R(\Delta C_R)_m \, a \cdot Z \qquad (9\text{-}14)$$

where $(\Delta C_R)_m$ is the logarithmic mean of $(C_R - C_R{}^*)_1$ and $(C_R - C_R{}^*)_2$, and K_R is the overall transfer coefficient based on concentration in the raffinate phase. $C_R{}^*$ is the concentration in the raffinate phase in equilibrium with concentration C_E in the extract phase.

Example 9-6: Acetic acid is extracted from an aqueous solution by benzene in a small packed column of $0\cdot004\,55$ m^2 cross-sectional area. The concentrations in each stream are as follows:

acid concentration in feed	$= 0\cdot689$ kmol/m^3
acid concentration in raffinate	$= 0\cdot683$ kmol/m^3
acid in inlet solvent	$= 0\cdot003\,97$ kmol/m^3
acid concentration in extract	$= 0\cdot011\,47$ kmol/m^3

The benzene passes up the column with a flow of $5\cdot67$ cm^3/s. If the ratio of the concentration of acid in benzene to that in water at equilibrium is given by $C_E{}^* = 0\cdot0247\,C_R$ and the overall mass transfer coefficient based on concentrations in the benzene phase is $K_E \cdot a = 0\cdot000\,778$ 1/s, calculate the height of packing required.

Solution For conditions in the extract phase, eqn. 9-14 becomes

$$L_E(C_{E_2} - C_{E_1}) = K_E \cdot a(\Delta C_E)_m \cdot Z \qquad (9\text{-}15)$$

In this case

$$L_E = (5\cdot67 \times 10^{-6}/0\cdot004\,55) = 1\cdot245 \times 10^{-3} \text{ m}^3/\text{m}^2 \text{ s}$$

then

$$(C_{E_2} - C_{E_1}) = (0\cdot011\,47 - 0\cdot003\,97) = 0\cdot007\,50 \text{ kmol/m}^3$$

At the bottom of the column, $C_{R_1} = 0\cdot683$ kmol/m^3, therefore

$$C_{E_1}^* = (0\cdot0247 \times 0\cdot683) = 0\cdot0169 \text{ kmol/m}^3$$

and

$$\Delta C_1 = (C_{E_1}^* - C_{E_1}) = (0 \cdot 0169 - 0 \cdot 003\ 97) = 0 \cdot 012\ 93\ \text{kmol/m}^3$$

At the top of the column, $C_{R_2} = 0 \cdot 689\ \text{kmol/m}^3$, therefore

$$C_{E_2}^* = (0 \cdot 0247 \times 0 \cdot 689) = 0 \cdot 0171\ \text{kmol/m}^3$$

and

$$\Delta C_2 = (C_{E_2^*} - C_{E_2}) = (0 \cdot 0171 - 0 \cdot 011\ 47) = 0 \cdot 005\ 63\ \text{kmol/m}^3$$

Hence

$$(\Delta C)_m = (\Delta C_1 - \Delta C_2)/\ln(\Delta C_1/\Delta C_2)$$
$$= (0 \cdot 012\ 93 - 0 \cdot 005\ 63)/2 \cdot 303\ \log_{10}(0 \cdot 012\ 93/0 \cdot 005\ 63)$$
$$= 0 \cdot 008\ 75\ \text{kmol/m}^3$$

Thus, in eqn. 9-15

$$(0 \cdot 001\ 245 \times 0 \cdot 007\ 50) = (0 \cdot 000\ 778 \times 0 \cdot 008\ 75\ Z)$$

from which

$$Z = 1 \cdot 37\ \text{m}$$

Example 9-7: In a pilot plant test, acetic acid is extracted from a dilute aqueous solution by means of a ketone in a small spray tower $0 \cdot 0467\ \text{m}$ diameter and effective height $1 \cdot 09\ \text{m}$, the aqueous phase being run in at the top. The ketone enters the column acid free at a rate of $0 \cdot 0013\ \text{m}^3/\text{m}^2\text{s}$ and leaves with an acid concentration of $0 \cdot 376\ \text{kmol/m}^3$. The concentration of the aqueous phase falls from $1 \cdot 185\ \text{kmol/m}^3$ to $0 \cdot 0817\ \text{kmol/m}^3$. If the equilibrium conditions are expressed by concentration of acid in the ketone phase is $0 \cdot 548$ times that in the aqueous phase, calculate the overall extraction coefficient based on concentrations in the ketone phase and the height of the corresponding overall transfer unit. Hence estimate the overall extraction coefficient based on concentrations in the aqueous phase.

Solution   The solvent flowrate $L_E = 0 \cdot 0013\ \text{m}^3/\text{m}^2$ s, therefore

$$(C_{E_2} - C_{E_1}) = 0 \cdot 376 - 0 = 0 \cdot 376\ \text{kmol/m}^3$$

At the bottom of the column $C_{R_1} = 0 \cdot 817\ \text{kmol/m}^3$, then

$$C_{E_1}^* = (0 \cdot 548 \times 0 \cdot 817) = 0 \cdot 448\ \text{kmol/m}^3$$

and

$$\Delta C_1 = (C_{E_1}^* - C_{E_1}) = 0 \cdot 448\ \text{kmol/m}^3$$

At the top of the column $C_{R_2} = 1 \cdot 185$ kmol/m^3, then

$$C_{E_2}^* = 0 \cdot 548 \times 1 \cdot 185 = 0 \cdot 649 \text{ kmol/m}^3$$

and

$$\Delta C_2 = (C_{E_2}^* - C_{E_2}) = (0 \cdot 649 - 0 \cdot 376) = 0 \cdot 276 \text{ kmol/m}^3$$

Therefore

$$(\Delta C)_m = (0 \cdot 448 - 0 \cdot 276)/2 \cdot 303 \log_{10}(0 \cdot 448/0 \cdot 276)$$
$$= 0 \cdot 355 \text{ kmol/m}^3$$

then

$$Z = 1 \cdot 09 \text{ m}$$

Therefore in eqn. 9-15

$$(0 \cdot 0013 \times 0 \cdot 376) = K_E a \times 0 \cdot 355 \times 1 \cdot 09$$

from which

$$K_E a = 0 \cdot 001\ 26\ 1/s$$

The height of an overall transfer unit, based on conditions in the extract phase is given by eqn. 9-11

$$(HTU)_{OE} = L_E/K_E a = 0 \cdot 0013/0 \cdot 001\ 26 = 1 \cdot 03 \text{ m}$$

From eqns 9-12 and 9-13,

$$(HTU)_{OR} = (HTU)_R + (L_R/mL_E)((HTU)_{OR} - (mL_E/L_R)(HTU)_R)$$
$$= (L_R/mL_E)(HTU)_{OE}$$

or, in terms of the transfer coefficients

$$L_R/K_R \cdot a = (L_R/mL_E)(L_E/K_E \cdot a)$$

from which

$$K_R \cdot a = mK_E a = (0 \cdot 548 \times 0 \cdot 001\ 26) = 0 \cdot 000\ 69\ 1/s$$

9.3.2 Column diameter

Where a dispersed phase is rising unhindered through a continuous phase, the column diameter may be estimated from the conditions at which the column floods by allowing a reasonable margin of safety. The superficial velocities, assuming the particular phase alone occupies the column, are $v_d = Ld/A$ (m/s) for the dispersed phase, and $v_c = Lc/A$

(m/s) for the continuous phase, where $A(m^2)$ is the column cross section and $L_d(m^3/s)$ and $L_c(m^3/s)$ are the volumetric throughputs of the dispersed and continuous phases respectively. The sum of these two velocities gives the velocity of the drop relative to the continuous phase; that is, $\bar{v}_o$, the characteristic droplet velocity.

$$(v_d/\phi) + v_c/(1 - \phi) = \bar{v}_0(1 - \phi) \tag{9-16}$$

where ϕ is the fractional hold-up of the dispersed phase in the column on a volume basis. The term $(1 - \phi)$ represents a correction factor for hindered rising effects. When the column floods, both $dv_d/d\phi$ and $dv_c/d\phi$ become zero, $\phi = \phi_f$ and

$$v_{d_f} = 2\bar{v}_0\phi_f^2 (1 - \phi_f) \tag{9-17}$$

whilst

$$v_{c_f} = \bar{v}_0(1 - \phi_f)^2 (1 - 2\phi_f) \tag{9-18}$$

Eliminating $\bar{v}_0$ from eqns. 9-17 and 9-18, and writing $Q = v_{d_f}/v_{c_f}$,

$$\phi_f = ((Q^2 + 8Q)^{0.5} - 3Q)/4(1 - Q) \tag{9-19}$$

This treatment holds for all devices except packed columns providing no coalescence of droplets occurs.

Example 9-8:   The following hold-up data have been obtained below the floodpoint in a small continuous counter-current contactor:

superficial velocity of dispersed phase						
v_d (mm/s)	0·71	1·27	1·95	2·41	2·88	3·39
superficial velocity of continuous phase						
v_c (mm/s)	0·36	0·64	0·97	1·20	1:44	0·68
fractional holdup						
ϕ	0·05	0·09	0·15	0·20	0:30	0·32

Assuming the droplet size is independent of the dispersed phase hold-up, calculate:
(a) the complete floodpoint curve for the system,
(b) the recommended column diameter for a continuous phase flowrate of $314·6$ cm^3/s and a dispersed phase flowrate of $196·7$ cm^3/s.

Solution   (a) In eqn. 9-16, when $\phi = 0$, $v_d = 0$, and hence v_d/ϕ is indeterminate. However multiplying throughout by ϕ gives

$$v_d + v_c\phi/(1 - \phi) = \bar{v}_0(1 - \phi) \cdot \phi$$

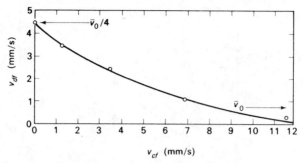

Fig. 9-8

and plotting $v_d + v_c\phi/(1 - \phi)$ against $(1 - \phi)\phi$ will give v_o as the slope of the function. The calculations are made as follows:

ϕ	0·05	0·09	0·15	0·20	0·30	0·32
$\phi(1 - \phi)$	0·0475	0·082	0·1275	0·160	0·210	0·217
$v_c\phi/(1 - \phi)$	0·019	0·063	0·171	0·300	0·616	0·320
$v_d + v_c\phi/(1 - \phi)$	0·729	1·333	2·121	2·710	3·496	3·710

These data are plotted in Fig. 9-8 from which the slope v_o is 17·86 mm/s.

For assumed values of ϕ_f, the superficial velocities at flooding may be obtained from eqns. 9-17 and 9-18 as follow:

ϕ	0·1	0·2	0·3	0·4	0·5	0·6
$\phi_f^2(1 - \phi_f)$	0·009	0·032	0·063	0·096	0·125	0·144
v_{df}	0·321	1·143	2·250	3·43	4·47	5·14
$(1 - \phi_f)^2$	0·81	0·64	0·49	0·36	0·25	0·16
$(1 - 2\phi_f)$	0·80	0·60	0·40	0·20	0	−0·20
v_{cf}	11·57	6·86	3·50	1·28	0	−0·571

The values of v_{df} and v_{cf} are plotted in Fig, 9-9, which represents the floodpoint curve for the system.

It should be noted that when $\phi_f = 0$, $v_{df} = 0$ from eqn. 9-17, and $v_{cf} = v_0$ from eqn. 9-18. Hence the floodpoint curve intersects with the v_{cf} axis at $v_{cf} = \bar{v}_0$. Similarly, when $\phi_f = 0·5$, $v_{cf} = 0$ from eqn. 9-18 and $v_{df} = v_0/4$ from eqn. 9-17. Thus the floodpoint curve meets the ordinate at $v_{df} = v_0/4$.

(b) From (a) $L_d = 196·7$ cm³/s, $L_c = 314·6$ cm³/s and $v_0 = 17·86$ mm/s from above.

The ratio $Q (= v_{df}/v_{cf})$, which approximates to L_{df}/L_{cf} is

$(196·7/314·6) = 0·625$.

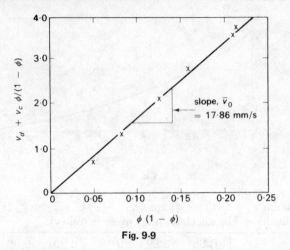

Fig. 9-9

Thus, in eqn. 9-19

$$\phi_f = ((0.625^2 + 8 \times 0.625)^{0.5} - (3 \times 0.625))/(4(1 - 0.625))$$
$$= 0.298$$

At flooding, in eqn. 9-17

$$v_{d_f} = L_d/A_f = 2\bar{v}_0 \, \phi_f^2 \, (1 - \phi_f)$$

therefore

$$A_f = L_d/(2\bar{v}_0 \phi_f^2 (1 - \phi_f)) = (196.7/(2 \times 1.79 \times 0.298^2 (1 - 0.298)$$
$$= 771.4 \text{ cm}^2 \text{ or } 0.077 \text{ m}^2$$

For design purposes, it is usual to work at 50% flooding, or $A = 2A_f$, therefore

$$A = 0.154 \text{ m}^2$$

and hence the design column diameter

$$D_c = (4 \times 0.154/\pi)^{0.5} = 0.45 \text{ m}$$

9.3.3 Packed columns

In packed columns with packings of greater than 12.5 mm diameter, a plot of v_d against hold up ϕ shows two transition points below the flood point. Below the lower transition point, droplets rise freely within the packing and v_d is a linear function of ϕ. Above this point, the rising of the droplets is hindered and the hold up increases very rapidly up to the second transition point. Above this, the hold-up becomes constant and v_d increases very rapidly to the flood point. During the constant hold-up regime, droplet coalescence occurs. For packed towers, eqn. 9-16 becomes

$$(v_d/\phi) + v_c/(1 - \phi) = \epsilon \bar{v}_0(1 - v) \tag{9-20}$$

where ϵ is the voidage of the packing. Equation 9-20 holds only for packings above a critical size, given by

$$d_{p_{crit}} = 2 \cdot 42 \, (\gamma/\Delta \rho g)^{0 \cdot 5} \text{ m} \tag{9-21}$$

where γ N/m is the interfacial tension, and $\Delta \rho$ kg/m^3 is the density difference between the continuous and the dispersed phases.

Example 9-9: A packed tower is to be designed to handle $0 \cdot 000 \, 35$ m^3/s of a hydrocarbon stream and $0 \cdot 000 \, 23$ m^3/s of an aqueous stream which is the continuous phase. The packing to be employed is $1 \cdot 25$ cm Raschig rings for which the following data are available: specific area, $a_p = 315$ m^2/m^3, voidage, $\epsilon = 0 \cdot 75$.

The density of the dispersed phase is 880 kg/m^3, the density of the continuous phase is 1000 kg/m^3 and the surface tension is $0 \cdot 030$ N/m. If the characteristic droplet velocity is $0 \cdot 025$ m/s, estimate:

    (a) the recommended tower diameter,
    (b) the fractional hold-up of the dispersed phase,
    (c) the Sauter mean droplet size,
    (d) the specific area of the phases, and
    (e) the minimum size of packing which may be employed.

Solution

(a) For operation under flooding conditions, the following correlation is applicable

$$1 + 0 \cdot 835(\rho_d/\rho_c)^{0 \cdot 25}(v_{d_f}/v_{c_f})^{0.5}$$
$$= C'((v_{c_f}^2 a_p/g\epsilon^3)(\rho_c/\Delta \rho)(\gamma/\gamma_w)^{0 \cdot 25})^{-0 \cdot 25} \tag{9-22}$$

where C' is a constant (= $0 \cdot 52$ for $12 \cdot 5$ mm Raschig rings), γ_w is the interfacial tension for air-water at 288 K (= $0 \cdot 073$ N/m), and $v_{c_f} = L_c/A_f$ ($2 \cdot 3 \times 10^{-4}/A_f$ m/s).

Thus, in eqn. 9-22

$$1 + 0 \cdot 835(880/1000)^{0 \cdot 25}(0 \cdot 000 \, 35/0 \cdot 000 \, 23)^{0 \cdot 5}$$
$$= 0 \cdot 52[(0 \cdot 000 \, 23)^2 \times 315/A_f^2 \times 9 \cdot 807$$
$$\times (0 \cdot 75)^3] \, [1000/(1000 - 880)] \, [(0 \cdot 030/0 \cdot 073)^{0 \cdot 25}]^{-0 \cdot 25}$$

from which

$$A_f = 0.0767 \text{ m}^2$$

Thus, designing for operation at 50% flooding,

$$A = 2A_f = 0.1534 \text{ m}^2$$

and the recommended diameter is

$$(4 \times 0.1534/\pi)^{0.5} = 0.442 \text{ m}$$

(b) For a tower of cross-sectional area 0.1534 m^2

$$v_c = L_c/A = 2.3 \times 10^{-4}/0.1534 = 1.5 \times 10^{-3} \text{ m/s}$$
$$v_d = L_d/A = 3.5 \times 10^{-4}/0.1534 = 2.28 \times 10^{-3} \text{ m/s}$$

Therefore in eqn. 9-20

$$(2.28 \times 10^{-3}/\phi) + (1.5 \times 10^{-3})/(1 - \phi)) = 0.75 \times 0.025(1 - \phi)$$

Solving by trial and error,

$$\phi = 0.595$$

(c) Above the critical packing size, the Sauter mean droplet diameter is given by

$$d_{vs} = 0.92(\gamma/\Delta\rho g)^{0.5}(\bar{v}_0 \epsilon \phi/v_d) \text{ m} \qquad (9\text{-}23)$$

then

$$d_{vs} = 0.92(0.03/(1000 - 880) \times 9.807)^{0.5}(0.025 \times 0.75$$
$$\times 0.595/2.28 \times 10^{-3}) = 2.27 \times 10^{-3} \text{ m or } 2.27 \text{ mm.}$$

(d) For unit volume of the column, volume of solvent held up is ϕ (m^3). Now, the volume of one drop is $\pi d^3/6$ (m^3) therefore number of drops is $6\phi/\pi d^3$ (1/m^3). The surface area of one drop is πd^2 (m^2), therefore total surface area is $(6\phi\pi d^2/\pi d^3) = 6\phi/d$ (m^2/m^3).
In the case of packed columns, an allowance must be made for the voidage, and hence, the specific area of the phases is

$$6\epsilon\phi/d = (6 \times 0.75 \times 0.595/2.27 \times 10^{-3}) = 118 \text{ m}^2/\text{m}^3$$

(e) The critical size of packing is obtained from eqn. 9-21:

$$d_{p \text{ crit}} = 2.42(0.030/(1000 - 880) \times 9.807)^{0.5} = 1.22 \times 10^{-2} \text{ m}$$
$$\text{or } 1.22 \text{ cm}$$

9.4 Leaching

Leaching is the extraction of a soluble component (which may be solid or liquid) from an insoluble solid by means of a solvent, and may be carried out continuously or batchwise. A leaching process will generally produce two output streams, one being simply a solution of solute and solvent, the other containing the inert solid mixed with some solute/solvent solution. The first stream can now be processed by a conventional separation technique, such as distillation, liquid–liquid extraction. The second stream can be further separated by a process known as *washing* using the same, or a different solvent. The washing process bears a great similarity to counter-current liquid-liquid extraction since one stage consists of an agitated vessel followed by a thickener, which is essentially the same as the mixer-settler considered earlier in this chapter,

9.4.1 Batch leaching and washing

Example 9-10:   500 kg of a solid A containing 28% by mass of a water soluble solid B is agitated in a vessel with 100 m³ water for 600 s. After decanting the solution, the residue is washed repeatedly in the same vessel with water and 25% of the solution produced remains in the residue after each washing. Water is saturated with B at a concentration of 2·5 kg/m³, and a pilot test in a tank of 1 m³ volume showed that 75% of saturation was attained in 10 s. Assuming conditions in the pilot unit represent those on the full-scale plant, what is the concentration of B in the solution decantered after leaching and how many washes are required after the initial leaching to reduce the concentration of B in A to 0·01% by mass.

Solution   The basic mass transfer equation for leaching takes the form

$$\mathrm{d}M/\mathrm{d}t = k'A(c_s - c)/b' \text{ kg/s}$$

where k' is the diffusion coefficient (m²/s), A the area of solid-liquid interface (m²), c the concentration of the solute in the bulk of the solution at time t (kg/m³), c_s the concentration of the saturated solution in contact with the solid (kg/m³), and b' the thickness of the liquid film adjacent to the solid (m).
For a batch process, where the total volume of solution V' is constant

$$\mathrm{d}M = V'\mathrm{d}c$$

265

therefore

$$dc/dt = k'A(c_s - c)/b'V'$$

Rearranging and integrating

$$\ln((c_s - c_0)/(c_s - c)) = k'At/V'b'$$

Where pure solvent is added initially $c_0 = 0$, and

$$c = c_s(1 - e^{-k'At/b'V'}) = c_s(1 - e^{-\alpha t/V'}) \text{ kg/m}^3 \qquad (9\text{-}24)$$

For the *pilot unit*, $c = (2\cdot5 \times 75/100) = 1\cdot875 \text{ kg/m}^3$, $c_s = 2\cdot5 \text{ kg/m}^3$, $V' = 1 \text{ m}^3$, and $t = 10 \text{ s}$.

Therefore, in eqn. 9-24

$$1\cdot875 = 2\cdot5(1 - e^{-10\alpha/1\cdot0})$$

from which

$$\alpha = 0\cdot138$$

For the *full scale unit*, $c_s = 2\cdot5 \text{ kg/m}^3$, $V' = 100 \text{ m}^3$ and $t = 600 \text{ s}$

and eqn. 9-24 becomes

$$c = 2\cdot5(1 - e^{-0\cdot138 \times 600/100}) = 1\cdot41 \text{ kg/m}^3$$

Therefore amount of B now in solution is

$$100 \times 1\cdot4 = 141 \text{ kg}$$

Amount of B initially as solid with A is

$$500 \times 28/100 = 140 \text{ kg}$$

and therefore all solid B has just gone into solution.

The liquor decantered will contain B with a concentration of $1\cdot4C \text{ kg/m}^3$

Considering the washing stage: if β is the amount of solvent decantered/amount of solvent remaining in the insoluble solid, then after one washing, the fraction of solute remaining with residue

$$\theta = 1/(1 + \beta)$$

Similarly, after λ washings, assuming β is constant

$$\theta = 1/(1 + \beta)^{\lambda} \qquad (9\text{-}25)$$

In the present case, it may be assumed that the amount of solution left in the residue after leaching is the same as that left after each washing, i.e. 25%. Hence, amount of B left in A after leaching is

$(100 \times 1{\cdot}40 \times 25/100) = 35$ kg

amount of B to be left in A after λ washings is

$(500 \times 0{\cdot}01/100) = 0{\cdot}05$ kg

Therefore fraction of B which remains after λ washings

$$\theta = (0{\cdot}05/35) = 0{\cdot}001\,43$$

and

$$\beta = (75/25) = 3{\cdot}0$$

Thus, in eqn. 9-25

$$0{\cdot}001\,43 = 1/(3 + 1)^{\lambda}$$

from which

$$\lambda = (\log_{10} 700/\log_{10} 4) = 4{\cdot}73$$

and hence five washes are required.

9.4.2 Continuous counter-current leaching

In this system similar to that shown in Fig. 9-2 (c), each residue is contacted with the extract from the next stage but one, and, although the mass balances are rather more complex, the conditions are not unlike those encountered in a distillation column where two counter-current flows are used to exchange components in a series of contacts. The treatment involves the use of triangular diagrams as the system corresponds to liquid-liquid extraction with partially miscible solvents. Although the inert solid is immiscible with the solvent, it does carry solvent and the associated dissolved solute in the residue to be processed in the next stage. This is akin to the condition of partial miscibility.

Example 9-11: $1{\cdot}60$ kg/s of a sand-salt mixture containing $62{\cdot}5\%$ sand is leached with $0{\cdot}5$ kg/s water in a counter-current unit. The residue from each stage contains $0{\cdot}25$ kg water/kg total solids. Assuming a stage efficiency of 75%, estimate the number of units to be included in the battery in order that the sand should contain 10% salt when dried. In the final residue it may be assumed that all the salt is in solution.

Solution   The solution is made graphically as shown in Fig. 9-10. Point F, representing the feed, $62{\cdot}5\%$ sand, $37{\cdot}5\%$ salt is marked on the graph.

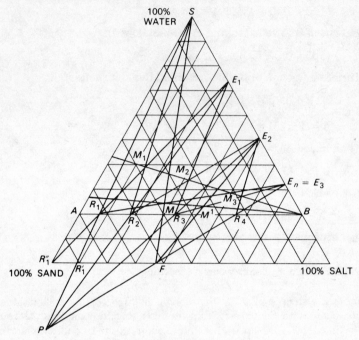

Fig. 9-10

The feed contains $62 \cdot 5\%$ sand, or $1 \cdot 60 \times 62 \cdot 5/100 = 1 \cdot 0$ kg/s. Assuming the dry solids are fed to the nth stage, the residue from the first stage, R_1 will also contain $1 \cdot 0$ kg/s sand. Where R_1 is dried, the salt concentration is to be 10%, or $0 \cdot 10$ kg/s salt. Indicating the dried residue as R'_1, the concentration of sand in this is

$$1 \cdot 0 \times 100/(1 \cdot 0 + 0 \cdot 10) = 90 \cdot 9\%$$

which is marked on the graph, and $R'_1 S$ drawn in. The mass of solvent/mass of total solids removed in each residue is $0 \cdot 25$ kg/kg, therefore

mass of solvent/(mass of solvent + mass of total solids) = $(0 \cdot 25/(1 + 0 \cdot 25)) = 0 \cdot 20$

If the solids were 100% sand, the residue would be represented by A, and similarly, for 100% salt by B. AB is drawn in and the intersection with SR'_1, represents the composition of the residue (wet) from the first stage, R_1. SF is now divided in the ratio $1 \cdot 60 : 0 \cdot 50$ (water:feed) giving point M, which is the mixture in the system taking the plant as a single unit. The intersection of $R_1 M$ produced to the water-salt axis, gives E_n the extract from the nth stage.

It may be assumed that in R_1, all the salt is in solution; the composition of this solution being E_1, the extract from the first stage. E_1 is therefore found by joining $R_1'' R_1$, where R_1'' represents 100% sand, and projecting to the water-salt axis. (If in R_1, some salt was undissolved, R_1'' would have been moved accordingly along $R_1'' R_1'$) SR_1', and $E_n F$ projected meet at the operating point P, as for liquid-liquid extraction in Example 9-5. Mass balances show that E_1 and R_2 lie on the same line through P and R_2 is the intersection of this line with AB because of the constant solvent/solids ratio in the residue. The point M_1 represents the composition of the mixture in stage 1, as this is the intersection of $E_1 R_1$ and SR_2. In this stage, and in every stage except the nth in which solid feed is introduced instead of wet residue, the ratio of 'free solvent' (i.e. not associated with solute) to inert solid is the same — this ratio is given by the intersection of BM_1 produced to the $R_1'' S$ axis. The composition of every mixture except that in the nth stage lies along the line $M_1 B$. E_2 is then located so that $E_2 R_2$ intersects with $E_1 R_3$ at M_2 which is on $M_1 B$. Proceeding in the same way, E_3 is found to coincide with E_n and hence three theoretical stages are indicated. In actual fact, for the 3rd = nth stage, $E_n R_3$ should intersect $E_2 F$ on the line MB. This is the case, the intersection being at M' and three theoretical stages will be specified. Therefore, actual stages required = $(3/0 \cdot 75) = 4$ stages.

9.4.3 Continuous counter-current washing

Where a solid solute has been leached, or in the cases of liquid solutes, we can assume that all the solute is in solution within the inert solid and this may then be removed by counter-current washing with the solvent. In this case, the solution of the problem is simpler, and two cases may be considered; constant underflow and variable underflow. In the former, the amount of solvent in the residue from each stage is constant, whereas with variable underflow, the amount is a function of the concentration of the solution.

Example 9-12: A vegetable seed material containing $0 \cdot 4$ kg oil/kg oily seeds is washed with a hydrocarbon solvent in order to recover the oil in a counter-current continuous unit equivalent to two theoretical stages. It is known from previous tests that $0 \cdot 05$ kg oil/kg oily seeds is within the structure of the seeds and is not removed by washing. If

0·43 kg solvent/kg oil-free seeds is removed in each residue and the extract from the second stage contains 50% oil, what is the composition of the residue leaving the plant and what will this be on an oil-free basis?

It is suggested that the underflow should not be constant and tests indicate that it will vary according to the concentration of the oil-solvent solution as shown in Table 9-3. If this is the case how many theoretical stages will be required to effect the same overall extraction?

Table 9-3   Variation of solution in underflow

concentration of solute in solution (kg solute/kg solution)	amount of solution in underflow (kg solution/kg oil-free seeds)
0	0·43
0·1	0·47
0·2	0·52
0·3	0·58
0·4	0·65
0·5	0·73
0·6	0·83
0·7	0·99
0·8	1·23
0·9	1·60
1·0	2·34

Solution
(a) *Constant underflow*

The solution is obtained graphically as shown in Fig. 9-11.

The feed is 0·40 kg oil/kg oily seeds or 40% oil, 60% seeds marked as point F. It is known that in addition to free-oil, the final residue when dried will contain 0·05 kg oil/kg oily seeds, i.e. 5% oil, 95% dry seeds shown by R_1'' on the diagram. Similarly, the extract from the second stage, $E_2 = E_n$ is fixed at 0·5 kg oil/kg solution. The underflow is given as 0·43 kg solvent/kg oil free seeds. Thus, in all oil-free mixtures,

% seeds = 100 x 1/(1 + 0·43) = 70%

which is marked as point A. For a constant ratio, as the amount of oil-free seeds tends to zero, the amount of solvent must also tend to

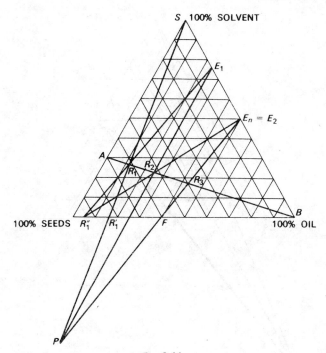

Fig. 9-11

zero and hence the AB line passes through the 100% oil apex. (This is as a result of the definition of underflow. If this had been given as kg solvent/kg oil + seeds, AB would have been parallel to the oil-seeds base line as in Example 9-11.) $E_2 F$ now cuts AB at R_3 and the operating point P lies on the projection of this line. Similarly $E_2 R_1''$ cuts AB at R_2, the residue from the second stage. The point P is now sought by trial and error until a position is found at which PR_2 projected meets $R_1'' R_1$ projected on the solvent-oil axis at E_1; R_1 being the intersection of $E_1 R_1''$ and PS. This is shown in Fig. 9-11 and the composition of the residue leaving the first stage, i.e. the unit is

$$R_1 = 61\cdot5\% \text{ seeds, } 27\cdot5\% \text{ solvent, } 11\cdot0\% \text{ oil}$$

The residue on a solvent-free basis is given by point R_1',

$$R_1', = 84\cdot0\% \text{ seeds, } 16\cdot0\% \text{ oil}$$

That is, on a solvent-free basis, the final residue contains $0\cdot16$ kg oil/kg oil seeds.

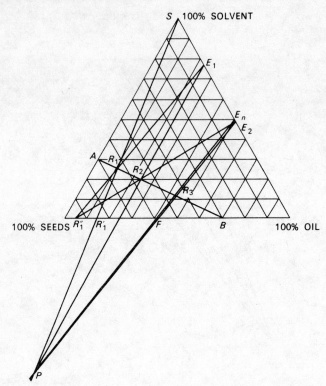

Fig. 9-12

(b) *Variable overflow*

The solution is shown in Fig. 9-12 where the points R_1'', R_1', F, E_n and P are as before. The important change is the line AB which may be plotted from the data given in Table 9-3. The composition of the solute in the solution in equilibrium with the solution in the underflow is calculated from the two values given plus the fact that the total mass fraction in any one mixture is unity. The line AB obtained in this way is drawn on the diagram.

Now PS intersects AB at R_1, and R_1'' R_1 produced meets the solvent-oil axis at E_1 because the residue R_1 is a mixture of the extract and the seeds containing bound oil. E_1P cuts AB at R_2, and R_1'' R_2 produced gives E_2 at the axis. This contains more oil than the desired extract E_n and hence two theoretical stages are more than adequate.

The variable underflow does not therefore greatly affect the operation although there is a slight margin of overdesign.

These rather simple graphical techiniques are designed to solve leaching and washing problems in terms of theoretical stages. This procedure may be carried out much more accurately by calculation which consists of a stagewise solution of the mass balances. Nevertheless, the graphical method enables very rapid trials of the performance of a system to be made (much quicker in fact than explaining how to do them), and, as a solution to the nearest whole stage is usually required, no great sacrifice is made as far as loss of accuracy is concerned.

List of symbols

A	cross-sectional area of column	(m^2)
A	area of solid-liquid interface	(m^2)
a	flowrate of component A	(kg/s)
a	specific area	(m^2/m^3)
a_p	specific area of packing material	(m^2/m^3)
b	flowrate of component B	(kg/s)
b'	thickness of liquid film adjacent to solid	(m)
C	concentration of solute in solution	$(kmol/m^3)$
C'	constant in eqn. 9-22	$(-)$
$(\Delta C)_m$	logarithmic mean concentration difference	$(kmol/m^3)$
c	concentration of solute in bulk of solution	(kg/m^3)
D_c	column diameter	(m)
d	droplet diameter	(m)
$d_{p\,crit}$	critical packing size	(m)
d_{vs}	Souter mean droplet diameter	(m)
E	composition of extract phase	$(-)$
e	flowrate of extract	(kg/s)
f	flowrate of feed stream	(kg/s)
g	gravitational constant	$(9 \cdot 807\ m/s^2)$
HTU	height of a transfer unit	(m)
K	overall coefficient of mass transfer	(m/s)
k	film coefficient of mass transfer	(m/s)

k'	diffusion coefficient	(m^2/s)
L	liquid flowrate	$(m^3/m^2\,s)$ or (m^3/s)
M	mass of solute	(kg)
m	slope of equilibrium line	(−)
N	rate of mass transfer	$(kmol/m^2\,s)$
n	number of stages	(−)
NTU	number of transfer units	(−)
P	difference in flowrates $(r_n - e_{n+1})$	(kg/s)
Q	ratio (v_{d_f}/v_{c_f})	(−)
R	composition of raffinate or residue	(−)
r	flowrate of raffinate	(kg/s)
s	flowrate of solvent	(kg/s)
t	time	(s)
V'	total volume of solution	(m^3)
v	superficial phase velocity	(m/s)
$\bar{v}_0$	characteristic droplet velocity	(m/s)
x	mass of solute/mass of solvent in raffinate	(kg/kg)
y	mass of solute/mass of solvent in extract	(kg/kg)
Z	effective height of column	(m)
α	$(k'A/b')$ in eqn. 9-24	(m^3/s)
β	ratio of solvent decantered/solvent in insoluble solid	(kg/kg)
ϵ	voidage of packing material	(m^3/m^3)
ρ	density	(kg/m^3)
λ	number of washings	(−)
ϕ	fractional hold-up of dispersed phase	(−)
θ	fraction of solute remaining in residue after washing	(−)
γ	interfacial tension	(N/m)

Subscripts

1, 2 etc.	referring to streams leaving stages, 1, 2 etc.
c	referring to the continuous phase
d	referring to the dispersed phase
E	referring to the extract stream
F	referring to the feed stream
f	referring to flooding conditions
i	referring to conditions at the interface
n	referring to the nth stage

O	referring to an overall parameter
o	referring to time zero, i.e. an initial condition
R	referring to the raffinate or residue
S	referring to the solvent stream
s	referring to the saturated condition
w	referring to water

Further Reading

COULSON, J. M., RICHARDSON, J. F., BACKHURST, J. R. & HARKER, J. H. *"Chemical Engineering"* Vol. 2. 3rd Ed. Pergamon Press London (1979).

DELL, F. R. & PRATT, H. R. C. Flooding Rates in Packed Columns *Trans. I. Chem. E.* **29** 89 (1951).

TREYBAL, R. E. *"Liquid Extraction"* McGraw-Hill New York (1957).

SHERWOOD, T. K. & PIGFORD, R. L. *"Absorption and Extraction"* 2nd edition McGraw-Hill New York (1952).

GAYLER, R. & PRATT, H. R. C. Symposium on liquid-liquid extraction *Trans. I. Chem. E.* **29** 110 (1951).

THORNTON, J. D. *Chem Eng. Sci.* **5** 201 (1956).

OLIVER, E. D. *"Diffusional Separation Processes"* Wiley New York (1966).

ALDERS, L. *"Liquid-Liquid Extraction"* Elsevier New York (1955).

DAVIES, J. T. & RIDEAL, E. K. *"Interfacial Phenomena"* Academic Press London (1961).

FRANCIS, A. W. *"Liquid-Liquid Equilibriums"* Interscience New York (1963).

HENLEY, E. J. & STAFFIN, H. K. *"Stagewise Process Design"* Wiley New York (1963).

SAWISTOWSKI, H. & SMITH, W. *"Mass Transfer Process Calculations"* Wiley New York (1963).

TREYBAL, R. E. *"Mass Transfer Operations"* 2nd edition McGraw-Hill New York (1968).

BLACKADDER, D. A. & NEDDERMAN, R. M. *"A Handbook of Unit Operations"* Academic Press London (1971)

HANSON. C. *"Recent Advances in Liquid-Liquid Extraction"* Pergamon Press Oxford (1971)

NULL, H. R. *"Phase Equilibrium in Process Design"* Wiley-Interscience New York (1970)

Proceedings of International Solvent Extraction Conference ISEC 71 *"Solvent Extraction"* 2 volumes. Society of Chemical Industry London 1971

Problem 9-1: An aqueous solution is to have the concentration of a solute reduced from $147 \cdot 5 \text{ kg/m}^3$ to $32 \cdot 0 \text{ kg/m}^3$ by counter-current extraction with a solvent initially containing $7 \cdot 2 \text{ kg/m}^3$ solute. A mixer-settler installation equivalent to six theoretical stages is available. If the solvents may be regarded as immiscible, what solvent/aqueous phase ratio should be employed and what is the composition of the final extract phase?

The equilibrium data for the system are shown in Table 9-4. What action should be taken to preserve the composition of the final raffinate phase if one of the mixer-settlers failed in operation?

$$(0 \cdot 44, 58 \cdot 0 \text{ kg/m}^3)$$

Table 9-4. Equilibrium data for Problem 9-1

concentration of solute in solvent phase (kg/m³)	concentration of solvent in aqueous phase (kg/m³)
5·0	12·3
10·0	26·4
15·0	39·9
20·0	48·0
25·0	56·7
30·0	65·6
35·0	74·5
40·0	82·5
45·0	90·4
50·0	97·5
55·0	105·5
60·0	112·0

Problem 9-2: A counter-current multiple contact installation, equivalent to four theoretical stages is to be used for extracting a feed containing 60% solute B in solvent A with a pure solvent, C. If the final raffinate on a solvent C-free basis is to contain 2% B, calculate:

(a) the required solvent-feed ratio, and
(b) the composition of the final extract phase on a solvent C-free basis.

The equilibrium data for the partially miscible system are shown in Table 9-5.

$$(0 \cdot 81; 10 \cdot 9\% \text{ A}, 89 \cdot 1\% \text{ B})$$

Table 9-5.   Equilibrium data for problem 9-2

Raffinate Phase (% by mass)			Extract Phase (% by mass)		
A	B	C	A	B	C
42·0	50·0	8·0	9·5	54·0	36·5
56·0	40·0	4·0	5·5	45·2	49·3
67·0	30·0	3·0	3·5	34·5	62·0
77·0	20·0	3·0	2·5	23·5	74·0
87·0	10·0	3·0	1·75	11·5	86·75
92·8	5·0	2·2	1·5	5·0	93·5

Problem 9-3:   A solute is extracted from a hydrocarbon solvent into an aqueous phase in a continuous column contactor where the characteristic droplet velocity is 185% greater than that of 0·0275 m/s in the solute-free system. If the contactor is to handle 330 cm³/s and 570 cm³/s of dispersed and continuous phase respectively and the effective height of the column is 6·7 m calculate:

(a) the recommended column diameter,
(b) the volume of the dispersed phase hold-up.

$$(0·31 \text{ m}; 0·034 \text{ m}^3)$$

Problem 9-4:   An extraction column is filled with packing of voidage 0·75 m³/m³. When the dispersed and continuous superficial velocities are 0·0076 m/s and 0·0025 m/s respectively, the fractional hold-up of the dispersed phase is 0·22. If the density difference between the phases is 114 kg/m³ and the interfacial tension is 0·035 N/m, calculate:

(a) the Sauter mean droplet diameter,
(b) the specific area between the phases, and
(c) the smallest size of packing which may be used in the column. $(0·002\ 26 \text{ m}; 438 \text{ m}^2/\text{m}^3; 0·0043 \text{ m})$

Problem 9-5:   Gypsum is produced from a wet process phosphoric acid plant at the rate of 1·175 kg/s contaminated with 1 kg weak phosphoric acid containing 0·55 kg acid/kg weak acid per kg dry gypsum. The gypsum is to be washed with 2·35 kg/s water so that it contains 0·1% acid when dried. How many stages will be required if

(a) the residue from each stage contains 0·5 kg water/kg dry gypsum, and

277

(b) if the amount of weak acid removed in the residue is given by:

$$w = 0 \cdot 2a + 0 \cdot 30 \text{ kg/kg} ?$$

where w — mass of weak acid/mass dry gypsum in the residue (kg/kg)
    a — concentration of acid in the weak acid (kg/kg).

(3;2)

10. Humidification and Drying

10.1 Humidification

It is often necessary to change the amount of vapour in a gas stream. If, for example, the moisture content of an air stream is to be reduced, partial condensation followed by removal of the condensed vapour must take place. In humidification operations, the vapour content of a gas is increased, and for the air-water system, moisture is transferred to the air in contact with the water and this is carried into the bulk of the air stream by diffusion. As a result of the evaporation, heat is lost by the water stream and the operation is used for water-cooling as in a cooling tower.

10.1.1 Definitions

Humidity, $\mathscr{H}$ kg/kg = mass of vapour associated with unit mass of dry gas

Humidity of saturated gas, $\mathscr{H}_0$ kg/kg = humidity of the gas when saturated with vapour at a given temperature. (The percentage humidity = $100\mathscr{H}/\mathscr{H}_0$)

Humid Heat, s kJ/kg K = heat required to raise unit mass of dry gas *and* its associated vapour through unit temperature difference at constant pressure.

Humid volume, v m^3/kg = volume occupied by unit mass of dry gas *and* its associated vapour (Saturated volume = humid volume of the saturated gas)

Dew point, T_dK = temperature at which the gas is just saturated with vapour. When the gas is cooled, the first drop of liquid is formed at this temperature.

Percentage relative humidity, RH = 100 (partial pressure of vapour in the gas)/(partial pressure of vapour in the saturated gas).

Example 10-1: The pressure and temperature of air in a room are $101 \cdot 3$ kN/m³ and 301 K respectively; the percentage relative humidity being 30%. If the partial pressure of water vapour when the air is saturated with water vapour at 301 K is $3 \cdot 8$ kN/m², calculate:

(a) the partial pressure of water vapour in the room and the dewpoint,

(b) the specific volumes of the air and water vapour,

(c) the humidity of the air, and

(d) the percentage humidity.

Solution

(a) The percentage relative humidity is defined as

$$RH = 100 p_w / p_{w_0} \qquad (10\text{-}1)$$

where p_w is the partial pressure of water vapour in the air (kN/m²), and p_{w_0} is the value of p_w when the air is saturated with water vapour (kN/m²)

Thus

$$p_w = RH p_{w_0} / 100 = (30 \times 3 \cdot 8 / 100) = 1 \cdot 14 \text{ kN/m}^2$$

From Table 10-1, water has a vapour pressure of $1 \cdot 14$ kN/m² at a saturation temperature of 282 K and this is therefore the dewpoint, i.e. T_d = 282 K.

(b) In unit volume of air, the mass of water vapour is given by

$$p_w M_w / RT \text{ kg/m}^3 \qquad (10\text{-}2)$$

where M_w is the molecular weight of water (kg/mol), and R the Universal Gas Constant ($8 \cdot 314$ kJ/mol K).

Similarly, the mass of the air is

$$(P - p_w) M_A / RT \qquad (10\text{-}3)$$

where P is the total pressure (kN/m²), and M_A = "molecular weight" of air (kg/kmol).

The specific volume of water vapour is thus

$$v_w = RT / p_w M_w = (8 \cdot 314 \times 301 (1 \cdot 14 \times 18)) = 121 \cdot 9 \text{ m}^3/\text{kg}$$

Table 10-1. Vapour Pressure of Water

Temperature (T K)	Vapour Pressure (p_w kN/m^2)
275	0·72
280	1·02
285	1·42
290	1·96
295	2·67
300	3·60
305	4·80
310	6·33
315	8·26
320	10·70
325	13·75
330	17·47

Similarly, the specific volume of the air is

$$v_A = RT/(P - p_w)M_A$$

Taking the mean 'molecular weight' of air as 29, (chapter 2), then

$$v_A = (8·314 \times 301/(101·3 - 1·14) \times 29) = 0·862 \text{ m}^3/\text{kg}.$$

(c) The humidity is given by combining eqns. 10-2 and 10-3 as

$$\mathcal{H} = (p_w M_w/RT)/((P - p_w)M_A/RT) = p_w M_w/(P - p_w)M_A$$

For the air/water system, $p_w \ll P$ and hence

$$\mathcal{H} = p_w (18p_w/(P \times 29)) = 0·621 p_w/P$$

therefore

$$\mathcal{H} = (0·621 \times 1·14/101·3) = 0·0069 \text{ kg/kg}$$

(d) From the previous section

$$\mathcal{H}_0 = p_{w_0} M_w/(P - p_{w_0})M_A \tag{10-4}$$

The % humidity is

$$100 \mathcal{H}/\mathcal{H}_0 = 100 (p_w M_w/(P - p_w)M_A)/(p_{w_0} M_w/(P - p_{w_0})M_A)$$
$$= ((P - p_{w_0})/(P - p_w)) \times (100 p_w/p_{w_0})$$
$$= ((P - p_{w_0})/(P - p_w)) \times \% RH$$
$$= (101·3 - 3·8)/(101·3 - 1·14) \times 30 = 29·2\%$$

(N.B. % humidity = % RH only when $(P - p_{w_0})/(P - p_w) = 1$)

281

Example 10-2: A mixture of benzene vapour and nitrogen has a relative humidity of 60% at 297 K; the total pressure being $102 \cdot 4 \text{ kN/m}^2$. If the mixture is cooled to 283 K, to what pressure must the mixture be compressed in order to recover 75% of the benzene?

The vapour pressure of benzene is $12 \cdot 2 \text{ kN/m}^2$ and $6 \cdot 05 \text{ kN/m}^2$ at 297 K and 283 K respectively.

Solution
From eqn. 10-1

$$p_w = p_{w_0} \, RH/100$$

and therefore, at 297 K

$$p_w = (12 \cdot 2 \times 60/100) = 7 \cdot 32 \text{ kN/m}^2$$

In the benzene–nitrogen mixture

mass of benzene $= p_w \, M_w/RT = (7 \cdot 32 \times 78)/(8 \cdot 314 \times 297) = 0 \cdot 231$ kg

mass of nitrogen $= (P - p_w) M_A/RT = (102 \cdot 4 - 7 \cdot 32)28/(8 \cdot 314$
$\times \, 297) = 1 \cdot 078$ kg

Therefore, humidity

$$\mathscr{H} = (0 \cdot 231/1 \cdot 078) = 0 \cdot 214 \text{ kg/kg}$$

In order to recover 75% of the benzene, this must be condensed out by increase in pressure and the humidity will be reduced to 25% of the initial value. As the vapour will be in contact with liquid benzene, the nitrogen will be saturated with benzene vapour and hence, at 283 K

$$\mathscr{H}_0 = (0 \cdot 214 \times 25/100) = 0 \cdot 0535 \text{ kg/kg}$$

The humidity of the saturated gas is given by eqn. 10-4

$$\mathscr{H}_0 = p_{w_0} \, M_w/(P - p_{w_0}) \, M_A$$

or

$$P = p_{w_0} \, (1 + M_w/\mathscr{H}_0 M_A)$$

that is

$$P = 6 \cdot 05 \, (1 + 78/(0 \cdot 0535 \times 28)) = 321 \cdot 1 \text{ kN/m}^2$$

10.1.2 Wet bulb temperature

When a stream of unsaturated gas is passed over the surface of a liquid, the humidity of the gas is increased owing to evaporation of the liquid.

The liquid is thus cooled and heat is transferred from the gas. At equilibrium, the heat from the gas equals the heat required for evaporation, and in this condition the liquid temperature is the wet bulb temperature.

The rate of heat transfer from the gas is

$$Q = h_c A \ (T - T_w) \ \ \text{W} \tag{10-5}$$

where h_c is the convective heat transfer coefficient ($\text{W/m}^2\text{K}$), A is the area of heat transfer (m^2), and T and T_w are the gas and liquid temperatures respectively (K).

If λ J/kg is the latent heat of vaporization of the liquid, then the rate of evaporation,

$$r = h_c A \ (T - T_w)/\lambda \ \ \text{kg/s} \tag{10-6}$$

The rate of evaporation also depends upon the rate of diffusion of vapour into the bulk gas stream and this is given by

$$r = h_d A M_w \ (p_{w_0} - p_w)/RT \ \ \text{kg/s} \tag{10-7}$$

where h_d is mass transfer coefficient (m/s); the driving force between the interface and the gas bulk being $(p_{w_0} - p_w) \ \text{kN/m}^2$. In terms of humidity, eqn. (10-7) becomes

$$r = h_d A M_w (\mathcal{H}_w - \mathcal{H}) p_A M_A/RT = h_d A \rho_A (\mathcal{H}_w - \mathcal{H})$$

where $\rho_A \ (kg/m^3)$ is the density of the gas at partial pressure, p_A, $\mathcal{H}_w$ (kg/kg) is the humidity of the gas saturated with vapour at the wet bulb temperature, T_w K.

Thus from eqns. (10-6) and (10-8)

$$(\mathcal{H}_w - \mathcal{H}) = (h_c/h_d \rho_A \lambda) \ (T - T_w) \tag{10-9}$$

The ratio h_c/h_d is independent of gas velocity below a value of 5 m/s and hence T_w depends only upon the temperature and humidity of the gas. $(h_c/h_d \rho_A)$ is approximately $1 \cdot 1 \times 10^3$ for air and $1 \cdot 7$ to $2 \cdot 2 \times 10^3$ for organic liquids.

Example 10-3:   Moist air at 300 K has a wet bulb temperature of 290 K. If the latent heat of vaporization of water at 290 K is 2458 kJ/kg, estimate the humidity of the air and the percentage relative humidity. The total pressure is $101 \cdot 3 \ \text{kN/m}^2$.

Solution   From Table 10-1, the partial pressure of water vapour at 290 K = $1 \cdot 96 \ \text{kN/m}^2$. The humidity of air saturated with water vapour at the wet bulb temperature is given by eqn 10-4

$$\mathcal{H}_w = p_{w_o}M_w/(P - p_{w_o})M_A = (1\cdot96 \times 18/(101\cdot3 - 1\cdot96)29$$
$$= 0\cdot0122 \text{ kg/kg dry air}$$

Therefore, in eqn. 10-9, taking $(h/h_d\rho_A)$ as $1\cdot09 \times 10^3$ and $\lambda = 2458 \times 10^3$ J/kg, $(0\cdot0122) -\mathcal{H}) = 1\cdot09/2458)(300 - 290)$

From which

$$\mathcal{H} = 0\cdot0078 \text{ kg/kg}$$

In eqn. 10-4

$$0\cdot0078 = (18\,p_w/(101\cdot3 - p_w)29)$$

therefore

$$p_w = 1\cdot257 \text{ kN/m}^2$$

At 300 K, $p_{w_0} = 3\cdot60$ kN/m^2, from Table 10-1, and hence

$$\% RH = 100\,p_w/p_{w_0} = (100 \times 1\cdot257/3\cdot60) = 34\cdot9\%$$

10.1.3 Adiabatic saturation temperature

When a gas is in contact with a liquid for a period sufficient for equilbrium to be established, the gas becomes saturated with vapour and both gas and liquid attain the same temperature. In an insulated system, the loss in total sensible heat is equal to the latent heat of the liquid evaporated, and the equilibrium temperature is known as the adiabatic saturation temperature. If the gas temperature falls from T to the adiabatic saturation value T_s, during which time the humidity increases from $\mathcal{H}$ to $\mathcal{H}_s$, ($\mathcal{H}_s$ being the saturation value at T_s), then, for unit mass of dry gas,

$$(T - T_s)s = (\mathcal{H}_s - \mathcal{H})\lambda$$

or

$$(\mathcal{H} - \mathcal{H}_s) = (- s/\lambda)(T - T_s) \qquad (10\text{-}10)$$

The humid heat, s, is virtually constant for small changes in humidity and examination of eqns. 10-9 and 10-10 shows that $s = h_c/h_d\rho_A$ when the adiabatic saturation and the wet bulb temperatures coincide. This is approximately true for the air-water system and accurately so when $\mathcal{H} = 0\cdot047$ kg/kg.

10.2 Temperature-humidity charts

On this chart, which is vital to all humidification and drying calculations, the following quantities are plotted against dry bulb temperature:

(a) Humidity, $\mathscr{H}$ for various values of the percentage relative humidity in the range 10–100%. The humidity at 100% *RH* is the saturation value given by eqn. 10-4.

(b) The specific volume of the dry gas, v_A which is a linear function of temperature.

(c) The humid volume of the saturated gas or saturated volume.

(d) The latent heat of vaporization of water.

In addition, the humid heat s is plotted as a function of humidity. Curves of humidity against temperature for gases with a given adiabatic saturation temperature are also included in the chart. These are known as adiabatic cooling lines and have a slope of $-s/\lambda$. For the air–water system, the adiabatic cooling lines represent conditions of constant wet bulb temperature and they enable the change in the vapour content of a gas stream to be evaluated as it is humidified by contact with water at the adiabatic saturation temperature of the gas. A simplified chart for the air–water system is included as Fig. 10-1.

Example 10-4: Air with a humidity of $0·005$ kg/kg dry air is heated to 325 K and passed to a tray dryer consisting of four sections in series. The air leaves the first section A with a relative humidity of 60% after which it is reheated to 325 K before entering the second section B which it also leaves at 60% RH. The same sequence is repeated in sections C and D and the moist air leaves the dryer at a flowrate of $4·75$ m^3/s. Neglecting heat losses and assuming that the wet material reaches the wet bulb temperature in each section, calculate

(a) the temperature of the wet material at the end of each section,

(b) the rate of water removal, and

(c) the required inlet air temperature if the drying is to be carried out in a single stage.

Solution   The calculation is made by means of the humidity chart, Fig. 10-1, and the relevant data are shown in Fig. 10-2.

(a) The air enters section A of the dryer at 325 K and a humidity of $0·005$ kg/kg. This is equivalent to a wet bulb temperature of 296 K. During drying it is assumed that the air and the wet material are both at

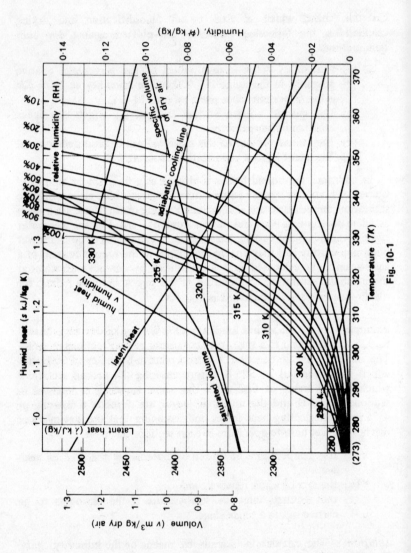

Fig. 10-1

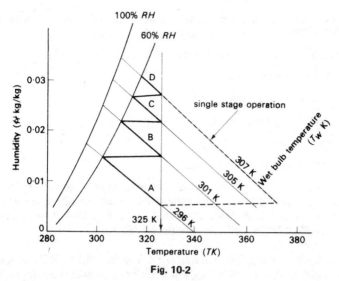

Fig. 10-2

this wet bulb temperature and hence humidification takes place along an adiabatic cooling line until $RH = 60\%$, when the air leaves section A. At this stage, examination of Fig. 10-1 shows that $T = 301$ K and the humidity has been increased to 0.015 kg/kg; the wet bulb temperature being the same as at the inlet, i.e. $T_w = 296$ K.

The air is now reheated to 325 K represented by a horizontal line on the chart, i.e. $\mathscr{H} = 0.015$ kg/kg, at which point $T_w = 301$ K. The sequence is now repeated and the data may be summarized as follows:

At the end of section A, $T = 301$ K, $T_w = 296$ K and $\mathscr{H} = 0.015$ kg/kg
At the end of section B, $T = 308$ K, $T_w = 301$ K and $\mathscr{H} = 0.022$ kg/kg
At the end of section C, $T = 312$ K, $T_w = 305$ K and $\mathscr{H} = 0.027$ kg/kg
At the end of section D, $T = 315$ K, $T_w = 307$ K and $\mathscr{H} = 0.032$ kg/kg.

The temperature of the wet material at the end of each section of the dryer is equal to the wet bulb temperature at that point, that is

A 296 K, B 301 K, C 305 K and D 307 K.

(b) The air leaves the dryer at 315 K and 60% *RH.*
From Fig. 10-1, the specific volume of dry air at 315 K is 0.893 m³/kg, and the saturated volume (i.e. 100% RH) at 315 K is 0.968 m³/kg.

Thus the humid volume of air at 60% RH and 315 K is approximately
$(0·893 + (0·968 − 0·893)60/100) =, \; 0·938 \, m^3/kg$

Therefore, mass of moist air leaving the dryer is

$(4·75/0·938) = 5·05 \, kg/s$

The total increase in humidity is

$(0·032 − 0·005) = 0·027$

and mass of water evaporated

$(5·06 \times 0·027) = 0·137 \, kg/s$

(c) For the drying to be carried out in a single stage (shown by the broken line in Fig, 10-2), the air must be heated initially to a wet bulb temperature of 307 K. With $\mathscr{H} = 0·005$ kg/kg, Fig. 10-1 shows that this corresponds to a dry bulb temperature of 380 K.

Exple 10-5: A fertilizer material is dried from 5·0% to 0·5% moisture (dry basis) at the rate of 1·5 kg/s in a rotary counter-current dryer. The solid enters at 294 K, leaving at 339 K and the air enters at 400 K ($\mathscr{H} = 0·007$ kg/kg), leaving at 355 K. Ignoring heat losses from the unit, estimate the throughput of dry air and the humidity of the air leaving the dryer. The specific heats of dry fertilizer and water vapour are 1·89 kJ/kg K and 2·01 kJ/kg K respectively.

Solution: The mass of dry air supplied is determined by means of a heat balance as follows

mass of water in inlet fertilizer	$= (1·5 \times 5/100)$
	$= 0·075 \, kg/s$
mass of water in outlet fertilizer	$= (1·5 \times 0·5/100)$
	$= 0·0075 \, kg/s$
therefore, rate of water evaporation	$= (0·075 − 0·0075)$
	$= 0·0675 \, kg/s$

Taking a datum temperature of 273 K

heat in wet fertilizer = heat in fertilizer + heat in moisture = $1·5 \times 1·89(294 − 273) + 0·075 \times 4·18(294 − 273)$ kJ/s $= 66·14 \, kW$

heat in 'dried' fertilizer = heat in fertilizer + heat in moisture = $1·5 \times 1·89(339 − 273) + 0·0075 \times 4·18(339 − 273)$ kJ/s $= 189·22 \, kW$

Heat gained by fertilizer, neglecting losses is

$(189 \cdot 22 - 66 \cdot 13) = 123 \cdot 09$ kW.

From Fig. 10-1, latent heat of vaporization of water, assuming the evaporation takes place at 294 K, the inlet temperature of the wet feed, is 2455 kJ/kg. Therefore

heat transferred to water evaporated

= (latent heat) + (sensible heat in vapour at the outlet)

= $(0 \cdot 0675 \times 2455) + (0 \cdot 0675 \times 2 \cdot 01(355 - 294))$ kJ/s

= $173 \cdot 98$ kW

Thus, heat lost by the inlet air stream is

$(173 \cdot 98 + 123 \cdot 09) = 297 \cdot 07$ kW.

As the evaporated moisture in the outlet air has already been considered, the inlet moist air may be considered separately. Assuming a' kg/s dry air are fed to the dryer, for $\mathscr{H} = 0 \cdot 007$ kg/kg, the humid heat = $1 \cdot 01$ kJ/kg K from Fig. 10-1.

Thus, heat lost by the inlet air stream is

$a' \times 1 \cdot 01(400 - 355) = 45 \cdot 45\, a'$ kW

Therefore, heat balance

$45 \cdot 45\, a' = 297 \cdot 07$

and

$a' = 6 \cdot 54$ kg/s

Moisture in the inlet air is

$0 \cdot 007$ kg/kg dry air = $(0 \cdot 007 \times 6 \cdot 54) = 0 \cdot 0457$ kg/s

and water evaporated = $0 \cdot 0675$ kg/s
Therefore, moisture in outlet air

$(0 \cdot 0675 + 0 \cdot 0457) = 0 \cdot 1132$ kg/s

and humidity of outlet air

$(0 \cdot 1132/6 \cdot 54) = 0 \cdot 0173$ kg/kg dry air

10.3 Cooling tower calculations

In a cooling tower, water is distributed over a packing of height z m, through which air passes upwards causing evaporation and hence cooling of the water stream. Referring to Fig. 10-3, G kg/m^2 s is the

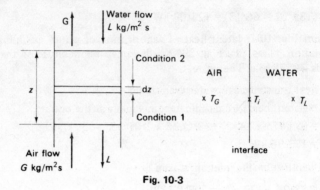

Fig. 10-3

mass flow of dry air and L the mass flow of water, usually assumed constant as the amount of evaporation is small. H is the enthalpy per unit mass of each stream. Several balances may now be made

(i) *Water balance*

$$dL = G \, d\mathcal{H} \tag{10-11}$$

(ii) *Enthalpy balance*
Over the packing height dz

$$G \, dH_G = L \, dH_L$$

but

$$H_L = s_L(T_L - T_0)$$

where s_L is the specific heat of the liquid (kJ/kg K), and T_0 the reference temperature (K).
Therefore

$$G \, dH_G = Ls_L \, dT_L$$

or over the entire column

$$G(H_{G_2} - H_{G_1}) = Ls_L(T_{L_2} - T_{L_1}) \tag{10-12}$$

(iii) *Heat transfer from the bulk liquid to the interface*

$$h_L a dz(T_L - T_i) = Ls_L \, dT_L$$

where h_L is the heat transfer coefficient in the liquid (kW/m² K), and a the interfacial area/unit volume of packing (m²/m³).
Therefore

$$dT_L/(T_L - T_i) = h_L a \, dz/(L \, s_L) \tag{10-13}$$

(iv) *Heat transfer from the interface to the bulk gas*

$$h_G \, a \, dz(T_i - T_G) = Gs \, dT_G$$

therefore

$$dT_G/(T_i - T_G) = h_G \, a \, dz/(G \, s) \qquad (10\text{-}14)$$

(v) *Mass transfer from the interface to the bulk gas*

$$h_{d}\rho_A a \, dz(\mathscr{H}_i - \mathscr{H}) = G \, d\mathscr{H}$$

therefore

$$d\mathscr{H}/(\mathscr{H}_i - \mathscr{H}) = h_{d}\rho_A \, a \, dz/G \qquad (10\text{-}15)$$

Since from eqn. 10-14, $h_G = h_{d}\rho_A$ s for the air-water system

$$G_s \, dT_G = h_G \, a \, dz(T_i - T_G) = h_{d}\rho_A \, a(sT_i - sT_G)dz \qquad (10\text{-}16)$$

Multiplying eqn. 10-15 by the latent heat λ,

$$G\lambda d\mathscr{H} = h_{d}\rho_A \, a(\lambda\mathscr{H}_i - \lambda\mathscr{H})dz \qquad (10\text{-}17)$$

Adding eqns. 10-16 and 10-17

$$G(sdT_G + \lambda d\mathscr{H}) = h_{d}\rho_A \, a \, dz(sT_i - \lambda\mathscr{H}_i - sT_G + \lambda\mathscr{H})$$

therefore

$$GdH_G = h_{d}\rho_A \, a \, dz(H_i - H_G) \qquad (10\text{-}17a)$$

Hence, integrating between the top and bottom of the column,

$$z = (G/(h_{d}\rho_A \, a))\int_1^2 dH_G/(H_i - H_G) \qquad (10\text{-}18)$$

Example 10-6: A small tower is to be designed to cool $0\cdot18$ kg/m²s water from 330 K to 290 K using a counter-current flow of air of $0\cdot6$ m³/m²s. The air enters at 290 K and has a relative humidity of 25%. Assuming that the whole of the resistance to mass and heat transfer is within the gas phase and that $h_d a = 0\cdot25$/s, estimate the height of packing required and the relative humidity of the exit air stream.

Solution   As the resistance to heat transfer lies entirely within the gas phase, it may be assumed that $T_i = T_L$ throughout the tower. The calculation is then made by evaluating the integral in eqn. 10-18 as shown in Fig. 10-4.

The first stage is to calculate the humidity of air saturated with water vapour as a function of temperature. The following equation may

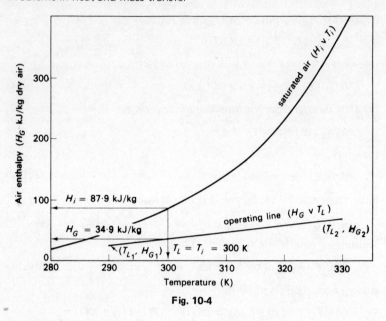

Fig. 10-4

be used:

$$H_G = s_A(T_G - T_0) + \mathcal{H}_0(s_V(T_G - T_0) + \lambda) \text{ kJ/kg dry air} \quad (10\text{-}19)$$

For example, at $T_G = 290$ K

$$H_G = 1 \cdot 01(290 - 273) + 0 \cdot 012(2 \cdot 01(290 - 273) + 2570)$$
$$= 47 \cdot 8 \text{ kJ/kg dry air}$$

where s_A and s_V are the specific heats of dry air and water vapour respectively (kJ/kg K) and $\mathcal{H}_0$ and λ are obtained from Fig. 10-1. Values of H_G at other temperatures are given in Table 10-2 and the data are plotted in Fig. 10-4. If it is assumed that air at the water surface is saturated with water vapour, then this curve represents a plot of H_i and T_i. The curve relating air enthalpy to water temperature, the operating line, may now be drawn in. This is obtained from eqn. 10-12

$$(H_{G_2} - H_{G_1}) = (Ls_L/G)(T_{L_2} - T_{L_1}) \quad (10\text{-}12)$$

Before evaluating the slope (Ls_L/G), it is necessary to obtain L in consistent units. The flow of air at the inlet is $0 \cdot 6 \text{ m}^3/\text{m}^2$ s. At 290 K and 25% RH, $\mathcal{H} = 0 \cdot 003$ kg/kg from Fig. 10-1. This is equivalent to $(0 \cdot 003 \times 29/18)$ kmol water/kmol dry air $= 0 \cdot 005 \text{ m}^3$ water/m^3 dry air.

Table 10-2   Enthalpy of Saturated Air

Temperature $(T_G$ K)	Saturated Humidity $(\mathscr{H}_0$ kg/kg dry air)	Enthalpy $(H_G$ kJ/kg dry air)
280	0·0060	19·7
285	0·0085	32·5
290	0·0120	47·8
295	0·0170	65·2
300	0·0225	87·9
305	0·0305	113
310	0·0410	143
315	0·0595	180
320	0·0725	230
325	0·0960	300
330	0·1280	385

Therefore, flowrate of dry air

$$G = 0·6 \times (1·0 - 0·005)(273/290)(29/22·4) = 0·727 \text{ kg/m}^2\text{s}$$

and slope,

$$Ls_L/G = (0·18 \times 4·18/0·727) = 1·035 \text{ kJ/kg K.}$$

The operating line passing through the point T_{L_1} (= 290 K) and H_{G_1}, which is obtained from eqn. 10-19

$$H_{G_1} = 1·01(290 - 273) + 0·003(2·01(290 - 273) + 2510)$$
$$= 24·80 \text{ kJ/kg dry air}$$

and may now be drawn on Fig. 10-4, noting that the top point is $T_{L_2} = 330$ K.

At T_{L_2} = 330 K, H_{G_2} = 65·6 kJ/kg dry air.
Combining eqns. 10-12, 10-13 and 10-17a gives

$$(H_G - H_i)/(T_L - T_i) = - h_L/h_d\rho_A \tag{10-20}$$

which gives the relationship between liquid temperature, air enthalpy and conditions at the interface at any point in the tower. This is represented by a series of lines of slope $- h_L/h_d\rho_A$. In the present case, it is assumed that there is no resistance to heat transfer within the liquid and hence the slope becomes $-\infty$, i.e. the lines joining equivalent conditions in both the gas and at the interface in Fig. 10-4 are parallel to the enthalpy axis.

Table 10-3 Determination of $1/(H_i - H_G)$

$T_L = T_i$ (K)	H_G (kJ/kg dry air)	H_i (kJ/kg dry air)	$(H_i - H_G)$ (kJ/kg dry air)	$1/(H_i - H_G)$ (kj/kg dry air)
290	24.8	47.8	23·0	0.0435
295	29.9	65·2	35.3	0.0284
300	34.9	87.9	53.0	0.0189
305	40.0	113	73.0	0.0137
310	45.1	143	97.9	0.0102
315	50.2	180	129.8	0.0077
320	55.2	230	174.8	0.0057
325	60.3	300	239.7	0.0042
330	65.6	385	319.4	0.0031

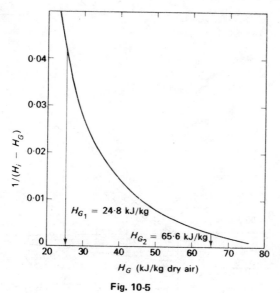

Fig. 10-5

Thus at $T_L = 300$ K, $H_G = 34\cdot9$ kJ/kg and since $- h_L/h_d\rho_A = -\infty$, $T_L = T_i$ and $H_i = 87\cdot99$ kJ/kg from the upper curve. Other values and the calculation of $1/(H_i - H_G)$ are shown in Table 10-3. Fig. 10-5 shows a plot of $1/(H_i - H_G)$ and H_G and the area under the curve between $H_{G_1} = 24\cdot8$ and $H_{G_2} = 65\cdot6$ kJ/kg is obtained as $0\cdot549$.

The density of the air is given by

$$\rho_A = (29/22\cdot4)(273/290) = 1\cdot219 \text{ kg/m}^3$$

and the height of the column is obtained from eqn. 10-18

$$z = (G/(h_d \, a\rho_A))\int_1^2 dH_G/(H_i - H_G)$$
$$= (0\cdot727/(0\cdot25 \times 1\cdot219)) \times 0\cdot549 = 1\cdot31 \text{ m}$$

The relationship between H_G and T_G is obtained by means of the graphical construction shown in Fig. 10-6. From this curve it is seen that at $H_{G_2} = 65\cdot6$ kJ/kg, $T_{G_2} = 296$ K. Thus in eqn. 10-19,

$$65\cdot6 = 1\cdot01(296 - 273) + \mathscr{H}_2 (2\cdot01(296 - 273) + 2510)$$

from which

$$\mathscr{H}_2 = 0\cdot0165 \text{ kg/kg}$$

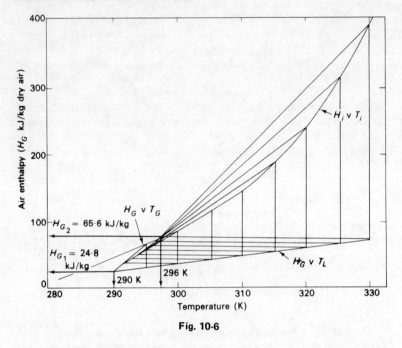

Fig. 10-6

At 296 K, and $\mathscr{H} = 0{\cdot}0165$, the relative humidity is obtained from Fig. 10-1 as $RH = 91\%$

10.4 Drying

As with humidification, the drying of solids involves simultaneous mass and heat transfer. The moisture content of a material, usually expressed as a percentage of the mass of dry material, varies when the material is exposed to air at a given temperature and humidity and water is gained or lost until an equlibrium moisture content is attained — this varies with both the material and the temperature. For example, a non-porous insoluble solid such as sand has an effective equilibrium moisture content of zero, whilst the equilibrium moisture content of a fibrous organic material may be quite high.

Example 10-7: 126 g/s of a product containing 4% water, is produced in a dryer from a wet feed containing 42% water (wet basis). Ambient air at 294 K and 40% *RH* is heated to 366 K in a preheater before entering the dryer which it leaves at 60% *RH*. Assuming the dryer operates adiabatically, calculate the amount of air supplied to the preheater and the heat required in the preheater. How will these values be affected if the air enters the dryer at 340 K and sufficient heat is supplied within the dryer so that the air leaves also at 340 K again with a relative humidity of 60%?

Solution   Referring to Fig. 10-1, air at 294 K and 40% *RH* has a humidity of 0·006 kg/kg. This remains unchanged on heating to 366 K and at the dryer inlet the air has a wet bulb temperature of 306 K. In the dryer, cooling takes place along the adiabatic cooling line until 60% *RH* is reached. At this point, the humidity is 0·028 and the dry bulb temperature 312 K. Thus the water picked up by 1 kg of dry air is

$$(0·028 - 0·006) = 0·22 \text{ kg water vapour}$$

The wet feed contains 42 kg water/100 kg feed, which is

$$42 \text{ kg water}/(100 - 42) \text{ kg dry solids} = 0·725 \text{ kg/kg dry solids.}$$

The product contains 4kg water/100 kg product, which is

$$4 \text{ kg water}/(100 - 4) \text{ kg dry solids} = 0·0417 \text{ kg/kg dry solids}$$

Therefore, the water evaporated is

$$(0·725 - 0·0417) = 0·6833 \text{ kg/kg dry solids}$$

Throughput of dry solids is

$$(126/1000)((100 - 4)/100) = 0·121 \text{ kg/s}$$

and evaporation is

$$(0·121 \times 0·6833) = 0·0825 \text{ kg/s}$$

Then, air throughput required is

$$0·0825/(0·028 - 0·006) = 3·76 \text{ kg/s}$$

Interpolation between the specific volume curves for dry and saturated air in Fig. 10-1 gives the specific volume at 294 K as 0·84 m³/kg. The volume of air required is therefore

$$(0·84 \times 3·76) = 3·16 \text{ m}^3/\text{s}$$

At $\mathscr{H} = 0·006$ kg/kg, the humid heat from Fig. 10-1 is 1·02 kJ/kg K and hence, heat required in the preheater is

$$3·76 \times 1·02(366 - 294) = 276 \text{ kW}$$

For the second case in which air both enters and leaves the dryer at 340 K, the outlet humidity is 0·111 kg/kg, therefore water picked up by the air is

$$(0.111 - 0.006) = 0.105 \text{ kg/kg dry air}$$

and the air requirements are

$$(0.0825/0.105) = 0.786 \text{ kg/s or } (0.786 \times 0.84) = 0.66 \text{ m}^3/\text{s}$$

As before, heat required in the preheater is

$$0.66 \times 1.02(340 - 294) = 31.0 \text{ kW}$$

[The heat to be supplied within the dryer is that required to heat the water to 340 K plus the latent heat at 340 K, that is

$$0.0825(4.18(340 - 294) + 2345) = 209 \text{ kW}$$

The total heat to be supplied is therefore 246 kW.]

10.4.1 The rate of drying

The capacity of a thermal dryer depends upon the rates of both heat and mass transfer. Heat is transferred by either direct contact between the material and hot gases as in a tunnel dryer, or by indirect heating where the solid comes into contact with steam or electrically heated surfaces. The shape of the drying rate curve, Fig. 10-7, depends upon the structure of the material involved and, in general, at least two regimes are apparent: the constant rate period (AB, DE) and the falling rate period (BC and, for sand, two such periods, EF and FC). E and B

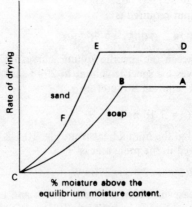

Fig. 10-7

are known as the critical moisture contents above which drying takes place from the saturated surface of the material by diffusion of water vapour through a stationary air film into the bulk air stream.

The mass rate of evaporation may be expressed as

$$r = K_G A(p_s - p_w) = K' A(p_s - p_w) u^{0.8} \text{ kg/s} \tag{10-20}$$

or, in terms of the heat transfer

$$r = h_c A \Delta T / \lambda \text{ kg/s} \tag{10-21}$$

where h_c, the heat transfer coefficient from air to the wet surface depends upon the air velocity and the direction of flow. During the constant rate period, the rate of drying depends upon h_c, ΔT and A, and is independent of conditions in the solid. In Fig. 10-7, the points B and E represent conditions where the surface is no longer capable of supplying sufficient free moisture to saturate the air which is in contact with the surface. The rate of drying during the *first falling rate* period depends upon the mechanism by which moisture from inside the material is transferred to the surface. At the end of this period, it is assumed that the surface is dry and the plane of separation moves into the solid. During the *second falling rate* period, evaporation takes place from within the solid with vapour reaching the surface by molecular diffusion. In this period, the rate of drying is independent of outside conditions.

10.4.2 Calculation of drying time

Assuming a material, whose surface is initially wet, is dried by a stream of hot air, the various moisture contents (dry basis) may be defined as shown in Fig. 10-8 which is a plot of the drying rate, R $(= (dw/dt)/A$ kg/m^2 s kg) and the moisture content w kg/kg dry solid.

(a) *constant rate period* In drying from w_0 to w_c, the surface is wet and the drying rate, R_c kg/m^2 s, is constant. The time taken to dry from w_0 to w_c, t_c is therefore

$$t_c = (w_0 - w_c)/R_c A \text{ s} \tag{10-22}$$

(b) *falling rate period* We can assume that the rate of drying is proportional to the free moisture content

$$f = (w - w_e)$$

299

f = free moisture content, $(w - w_e)$

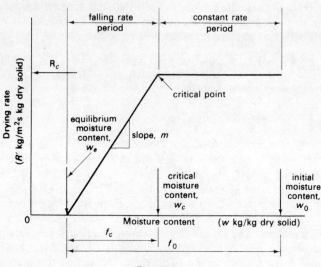

Fig. 10-8

or

$$- (dw/dt)/A = m(w - w_e) = mf$$

where m is the slope of the drying rate curve between w_c and w_e. Thus,

$$- 1/mA \int_{w_c}^{w} dw/(w - w_e) = \int_{0}^{t_f} dt$$

or

$$t_f = (1/mA)\ln((w_c - w_e)/(w - w_e)) = (1/mA)\ln(f_c/f) \text{ s}$$

$$(10\text{-}23)$$

where $f_c = (w_c - w_e)$, and hence $R_c = mf_c$

(c) *total time for drying*
This is given by

$$(t_c + t_f) = (w_0 - w_c)/mf_c\, A + (1/mA)\ln(f_c/f)$$
$$= (1/mA)\, ((f_0 - f_c)/f_c + \ln(f_c/f)) \text{ s} \qquad (10\text{-}24)$$

where $f_0 = (w_0 - w_e)$

Example 10-8: A wet solid is dried from 40% moisture to 8% in 20 ks. How long will it take to dry the material to 5% moisture under the same drying conditions? The critical moisture content is 15% and the equilibrium moisture content is 4%; all on a dry basis.

Solution   In this example

the initial moisture content, $w_0 = 0 \cdot 40$ kg/kg dry solid
the critical moisture content, $w_c = 0 \cdot 15$ kg/kg dry solid
the equilibrium moisture content, $w_e = 0 \cdot 04$ kg/kg dry solid

therefore

$$f_0 = (w_0 - w_e) = 0 \cdot 36 \text{ kg/kg}$$
$$f_c = (w_c - w_e) = 0 \cdot 11 \text{ kg/kg}$$

For the first set of conditions, $w = 0 \cdot 08$ kg/kg and $(t_c + t_f) = 20$ ks and $f = (w - w_e) = 0 \cdot 04$ kg/kg

Therefore, substituting in eqn. 10-24

$$20 = (1/mA)((0 \cdot 36 - 0 \cdot 11)/0 \cdot 11 + 2 \cdot 303 \log_{10}(0 \cdot 11/0 \cdot 04))$$

from which

$$mA = (3 \cdot 28/20) = 0 \cdot 164 \text{ kg/ks.}$$

For the second set of conditions, $w = 0 \cdot 05$ kg/kg and hence

$$f = (0 \cdot 05 - 0 \cdot 04) = 0 \cdot 01 \text{ kg/kg.}$$

mA and the other moisture contents are as before.

Hence, in eqn. 10-24,

$$(t_c + t_f) = (1/0 \cdot 164)(0 \cdot 36 - 0 \cdot 11)/0 \cdot 11 + 2 \cdot 303 \log_{10}(0 \cdot 11/0 \cdot 01))$$
$$= 28 \cdot 5 \text{ ks}$$

Example 10-9: An inorganic pigment is to be dried from 1 kg water/kg dry stock to $0 \cdot 01$ kg water/kg dry stock in a tray dryer consisting of a single tier of 50 trays each, $0 \cdot 02$ m deep and $0 \cdot 7$ m square completely filled with wet material. The mean air temperature is 350 K and the relative humidity across the trays can be considered constant at 10%. The mean air velocity is $2 \cdot 0$ m/s and the convective coefficient of heat transfer is given by

$$h_c = 14 \cdot 3 \ G^{0 \cdot 8} \text{ W/m}^2 \text{K}$$

where G is the mass velocity of dry air (kg/m^{2}s). The critical and equilibrium moisture contents of the pigment are $0 \cdot 3$ and 0 kg/kg dry stock respectively, and the density of the dry pigment is 600 kg/m^3. Assuming that drying is by convection from the top surface of the trays only, estimate the time of drying required.

Solution　Interpolation of Fig. 10-1, gives the specific volume of moist air at 350 K and $10\% RH = 1.06 \text{ m}^3/\text{kg}$.
The mass velocity of the air is therefore

$$G = (2.0/1.06) = 1.88 \text{ kg/m}^2 \text{s}$$

The convective heat transfer coefficient is then

$$h_c = (14.3 \times 1.88^{0.8}) = 23.8 \text{ W/m}^2\text{K or } 0.0238 \text{ kW/m}^2\text{K}$$

Assume that the temperature of the surface is equal to the wet bulb temperature of the air, thus is 317 K from Fig. 10-1. The mean temperature driving force is

$$\Delta T = (350 - 317) = 33 \text{ K}$$

The area for heat transfer, considering convection to the top surface only

$$A = (50 \times 0.7 \times 0.7) = 24.5 \text{ m}^2$$

At 317 K, the latent heat of vaporization of water, $\lambda = 2395 \text{ kJ/kg}$ from Fig. 10-1, and hence in eqn. 10-21, the rate of evaporation during the constant rate period

$$r = (0.0238 \times 24.5 \times 33/2395) = 0.008 \text{ } 05 \text{ kg/s}$$

The total volume of dry material is

$$(50 \times 0.7 \times 0.7 \times 0.02) = 0.49 \text{ m}^3$$

and hence, the mass of dry material is

$$(0.49 \times 600) = 294 \text{ kg}$$

The rate of drying during the constant rate period, R_c is then

$$R_c = 0.008 \text{ } 05/(294 \times 24.5) = 1.12 \times 10^{-6} \text{ kg water/m}^2 \text{s kg dry solid}$$

In this problem, the initial moisture content $w_0 = 1.0 \text{ kg/kg}$ dry solid, the critical moisture content $w_c = 0.3 \text{ kg/kg}$ dry solid, the equilibrium moisture content $w_e = 0$, and the final moisture content $w = 0.01 \text{ kg/kg}$ dry solid. Therefore $f_0 = 1.0, f_c = 0.3$ and $f = 0.01 \text{ kg/kg}$ dry solid.

As apparent from Fig. 10-8

$$R_c = mf_c$$

therefore

$$m = (1 \cdot 12 \times 10^{-6}/0 \cdot 3) = 3 \cdot 73 \times 10^{-6}/\text{m}^2\,\text{s}$$

Hence, in eqn, 10-24, the total time for drying

$$(t_c + t_f) = [1/(3 \cdot 73 \times 10^{-6} \times 24 \cdot 5)][(1 \cdot 0 - 0 \cdot 3)/0 \cdot 3$$
$$+ 2 \cdot 303 \log_{10}(0 \cdot 3/0 \cdot 01)] = 614\text{s}$$

Example 10-10:   Skeins of a synthetic fibre are dried from 46% to 8·5% moisture (wet basis) in a tunnel dryer 10 m long by a counter-current flow of hot air. The mass velocity of the air is 1·36 kg/m²s and at the inlet the temperature and humidity are 355 K and 0·03 kg/kg respectively. The air temperature is maintained at 355 K throughout the dryer by internal heating and the outlet humidity is 0·08 kg/kg dry air. The equilibrium moisture content is 0·25 × (%RH of the air) and the drying rate is given by

$$R' = 1 \cdot 34 \times 10^{-4} G^{1 \cdot 47}(w - w_e)/(\mathscr{H}_w - \mathscr{H})\ \text{kg/kg dry fibres}$$

where $\mathscr{H}_w$ is the saturation humidity at the wet bulb temperature.

Data relating w, $\mathscr{H}$ and $\mathscr{H}_w$ are shown in Table 10-4. At what velocity should the skeins be passed through the dryer?

Table 10-4 Data for Example 10-10

w (kg/kg dry fibre)	$\mathscr{H}$ (kg/kg dry air)	$\mathscr{H}_w$ (kg/kg dry air)	% RH
0·852	0·080	0·095	22·4
0·80	0·0767	0·092	21·5
0·60	0·0635	0·079	18·2
0·40	0·0503	0·068	14·6
0·20	0·0371	0·055	11·1
0·093	0·030	0·049	9·0

Solution: The mass velocity of the air G is 1·36 kg/m² s. The rate of drying is thus

$$R' = 1 \cdot 34 \times 10^{-4} \times 1 \cdot 36^{1 \cdot 47}(w - w_e)/(\mathscr{H}_w - \mathscr{H})$$
$$= -2 \cdot 11 \times 10^{-4}(w - w_e)/(\mathscr{H}_w - \mathscr{H})\ \text{kg/s kg dry fibre}$$

The procedure is now to estimate the rate of drying using this relationship at various points in the dryer and this is best laid out in tabular form as shown in Table 10-5.

Table 10·5 Calculation of Drying Rate Curve

w (kg/kg dry fibre)	0·852	0·80	0·60	0·40	0·20	0·093
%RH	22·4	21·5	18·2	14·6	11·1	9·0
w_e (kg/kg dry fibre) $= 0·25\,RH/100$	0·056	0·054	0·046	0·037	0·028	0·023
$(w - w_e)$ (kg/kg dry fibre)	0·794	0·746	0·554	0·363	0·172	0·070
$\mathscr{H}_w$ (kg/kg dry air)	0·095	0·092	0·079	0·068	0·055	0·049
$\mathscr{H}$ (kg/kg dry air)	0·0800	0·0767	0·0635	0·0503	0·0371	0·0190
$(\mathscr{H}_w - \mathscr{H})$	0·015	0·0153	0·0155	0·0177	0·0179	0·0190
R' (kg/kg dry fibre s) $[= -2·11 \times 10^{-4}(w - w_e)/(\mathscr{H}_w - \mathscr{H})]$	−0·0112	−0·0103	−0·0076	−0·0043	−0·0020	−0·0008
$1/R'$ (kg dry fibre. s/kg)	−89·0	−97·0	−132	−233	−500	−1250

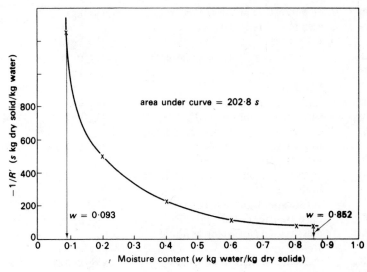

Fig. 10-9

An inlet moisture of 46·0% on a wet basis is equivalent to an initial moisture content of $46·0/(100 − 46·0) = 0·852$ kg/kg dry fibre. Similarly, the outlet moisture 8·5% is equivalent to $8·5/(100 − 8·5) = 0·093$ kg/kg dry fibre; the values shown in Table 10-4.

Table 10-5 shows the drying rate R as a function of the moisture content on a dry basis. The units of R' are kg water/s kg dry fibre and if this is divided into the moisture the units are s. In other words, the area under a plot of $1/R'$ and w is the drying time. Such a plot is shown in Fig. 10-9, from which the area between $w = 0·852$ and $w = 0·093$ is obtained as equivalent to 203 s.

The velocity at which material should be passed through the dryer is therefore $(10/203) = 0·0494$, say 0·05 m/s

List of symbols

A	area for heat and mass transfer	(m^2)
a	interfacial area/unit volume of packing	(m^2/m^3)
a'	throughput of dry air in Example 10-5	(kg/s)
f	free moisture content above equilibrium value	(kg/kg)

f_c	free moisture at the critical point	(kg/kg)
f_0	free moisture under initial conditions	(kg/kg)
G	mass velocity of air or gas stream	(kg/m²s)
H	enthalpy/unit mass	(kJ/kg)
$\mathcal{H}$	humidity	(kg/kg)
$\mathcal{H}_s$	saturation humidity at T_s K	(kg/kg)
$\mathcal{H}_w$	saturation humidity at T_w K	(kg/kg)
h_c	convective heat transfer coefficient in the gas	(W/m²K)
h_d	mass transfer coefficient	(m/s)
h_L	convective heat transfer coefficient in the liquid	(W/m²K)
K'	constant in Equation 10-20	(−)
K_G	mass transfer coefficient	(kg/m²s(kN/m²))
L	mass velocity of water or liquid stream	(kg/m²s)
M_A	molecular weight of air or gas	(kg/kmol)
M_w	molecular weight of water or liquid	(kg/kmol)
m	slope of drying rate curve	(1/m²s)
P	total air or gas pressure	(kN/m²)
p_A	partial pressure of air or gas	(kN/m²)
p_s	vapour pressure of water	(kN/m²)
p_w	partial pressure of water vapour	(kN/m²)
Q	rate of heat transfer	(W or kW)
R	universal Gas Constant	(8·314 kJ/kmol K)
R'	drying rate	(1/m² s)
R_c	drying rate at the critical moisture content	(1/m² s)
RH	percentage relative humidity	(−)
r	mass rate of evaporation	(kg/s)
s	humid heat	(kJ/kg K)
s_A	specific heat of air or gas	(kJ/kg K)
s_V	specific heat of vapour	(kJ/kg K)
T	temperature	(K)
T_d	dew point of vapour	(K)
T_0	reference temperature	(K)
T_s	adiabatic saturation temperature	(K)
ΔT	mean temperature difference	(K)
t_c	time for drying during constant rate period	(s)
t_f	time for drying during falling rate period	(s)

u	mean air velocity	(m/s)
v	humid volume	(m^3/kg)
v_A	specific volume of air or gas	(m^3/kg)
v_w	specific volume of vapour	(m^3/kg)
w	mass of moisture/unit mass dry solids	(kg/kg)
w_c	critical moisture content	(kg/kg)
w_e	equilibrium moisture content	(kg/kg)
w_o	initial moisture content	(kg/kg)
z	height of packing	(m)
ρ_A	density of air or gas	$(kg/m^3))$
λ	latent heat of vaporization	(kJ/kg)

Subscripts

G	referring to gas
i	referring to conditions at the interface
L	referring to liquid
0	referring to the saturated condition

Further reading

BOSWORTH, R. C. L. *"Transport Processes in Applied Chemistry"* Pitman London (1958).

DAVIES, J. T. & RIDEAL, E. K. *"Interfacial Phenomena"* Academic Press London (1961).

FORD, R. *"Drying"* Institute of Ceramics Textbook Series *3* Maclaren London (1960).

HENLEY, E. J. & STAFFIN, H. K. *"Stagewise Process Design"* Wiley New York (1963).

JACKSON, J. *"Cooling Towers"* Butterworths London (1951).

JOSELIN, E. L. *"Ventilation and Air Conditioning"* Arnold London (1934).

NONHEBEL, G. & MOSS, A. A. H. *"Drying of Solids in the Chemical Industry"* Butterworths London (1971).

NORMAN, W. S. *"Absorption, Distillation and Cooling Towers"* Longmans London (1962).

PENMAN, H. L. *"Humidity"* Inst. Physics Chapman London (1962).

SAWISTOWSKI, H. & SMITH, W. *"Mass Transfer Process Calculations"* Wiley New York (1963).

TREYBAL, R. E. *"Mass Transfer Operations"* 2nd edition McGraw-Hill New York (1968).

WILLIAMS, G. A. *"Industrial Drying"* Leonard Hill London (1971).

COULSON, J. M., RICHARDSON, J. F., BACKHURST, J. R. & HARKER, J. H. *"Chemical Engineering"* Vols 1, 2, 4 & 5, Pergamon Press London (1954).

ECKERT, E. R. G. *"Introduction to the Transfer of Mass and Heat"* McGraw-Hill New York (1950).

PERRY, J. H. *"Chemical Engineers Handbook"* 4th edition McGraw-Hill New York (1963).

BACKHURST, J. R. & HARKER, J. H. *"Process Plant Design"* Heinemann Educational Books London (1973)

PINKAVA, J. *"Unit Operations in the Laboratory"* Iliffe Books London (1970)

Problem 10-1: 500 000 m^3/day of gas is to be dried from a dew point of 294 K to a dew point of 277·5 K. If the vapour pressure of water is 2·5 kN/m^2 and 0·85 kN/m^2 at 294 K and 277·5 K respectively, how much water must be removed and what is the volume of gas after drying?

$$(0·0712 \text{ kg/s} \equiv 6150 \text{ kg/day}; 466 \text{ } 270 \text{ } m^3/\text{day})$$

Problem 10-2: 0·15 kg/s of a wet material is to be dried from 70% to 5% moisture (wet basis) in a counter-current dryer using air at 373 K containing water vapour equivalent to a partial pressure of 1·03 kN/m^2. The air leaves the dryer at 313 K (vapour pressure of water = 7·37 kN/m^2) and 70% RH. How much air is required to remove the moisture?

$$(4·04 \text{ kg/s})$$

Problem 10-3: Water is to be cooled in a small tower from 328 K to 292 K by means of air entering the bottom of the tower at 293 K with a relative humidity of 20%. The flow of water is 0·26 $kg/m^2 s$ and the flow of air is 0·68 $m^3/m^2 s$. Assuming $h_d a = 0·2/s$ and that resistance to mass and heat transfer lies within the gas phase, calculate the height of packing required.

$$(1·55 \text{ m})$$

Problem 10-4: A wet solid is dried from 35% to 10% moisture (wet basis) under constant drying conditions in 18 ks. The critical moisture content is 14% and the equilibrium moisture content is 4% (wet basis). Assuming that the drying rate is directly proportional to the free moisture content during the falling rate period, how long will it take to dry the solid from 35% to 6% moisture (wet basis)?

$$(23·4 \text{ ks})$$

Problem 10-5: A slab of a wet material containing 50% moisture (wet basis) has a mass of 4·54 kg dry. The equilibrium moisture content is

5% of the total mass when in contact with air at 294 K and 20% *RH* and the slab is 0·61 m x 0·91 m x 76 mm thick. The drying rate with a fixed air velocity assuming drying takes place from one face only is shown in Table 10-6.

How long will it take to dry the slab to 15% moisture (wet basis)?

(1·73 ks)

Table 10-6. Drying Rate Data (air at 294 K and 20% *RH*)

mass of wet slab	drying rate
(kg)	(kg/m^2 s)
9·07	0·0013
7·17	0·0013
5·26	0·001 17
4·22	0·001 04
3·27	0·000 91
2·77	0·000 52
2·49	0·000 26

Problem 10-6: A co-current adiabatic tunnel dryer is to be designed to dry 0·25 kg/s wet sand from 1·0 to 0·1 kg/kg dry solid using air with a dry bulb temperature of 366 K and a wet bulb temperature of 307 K. The sand enters and leaves the dryer at 307 K and the air leaves at 311 K.

In a laboratory test, using the above air conditions, the critical moisture content of the sand was 0·5 kg/kg dry sand and the equilibrium moisture content was zero. Above the critical point, the drying rate was 0·0013 kg/m^{2}s and below this point, the drying rate fell linearly to zero at zero moisture content.

Assuming the dryer holds 30 kg dry sand per m length and the effective area for drying is 0·061 m^2/kg dry sand, estimate the length of dryer required.

(183 m)

Problem 10-7: It is intended to dry 0·062 kg/s of a dry solid from 1·56 kg/kg dry solid to 0·1 kg/kg solid in a continuous counter-current dryer; the equilibrium moisture content being 0·05 kg/kg. The air flow will be 3·2 kg/s and it will enter the dryer at 333 K with a humidity of 0·0155 kg/kg. The air temperature is maintained at 333 K by steam coils within the dryer 25% by mass of the exit air (dry) is recirculated to the inlet.

309

The drying rate for the material is given by the equation:

$$d_w/dt = 0 \cdot 027(w - w_e)\Delta \mathscr{H}$$

where w is the total moisture content (kg/kg dry solid), w_e the equilibrium moisture content (kg/kg dry solid), t the time (s), and ΔH (the humidity of air in equilibrium with moisture in solid — humidity of the air (kg/kg dry air)). If the dryer will hold 14·9 kg/m length, how long should the dryer be?

(46·3 m)